Galaxies: structure and evolution

Galaxies: structure and evolution

Roger J. Tayler

Professor of Astronomy
and Director of Astronomy Centre
University of Sussex

CAMBRIDGE
UNIVERSITY PRESS

Published by the Press Syndicate of the University of Cambridge
The Pitt Building, Trumpington Street, Cambridge CB2 1RP
40 West 20th Street, New York, NY 10011-4211, USA
10 Stamford Road, Oakleigh, Melbourne 3166, Australia

First published by Wykeham Publications 1978
This revised edition first published by Cambridge University Press 1993

Printed in Great Britain at the University Press, Cambridge

A catalogue record for this book is available from the British Library

Library of Congress cataloguing in publication data

Tayler, R. J. (Roger John)
Galaxies: structure and evolution / Roger J. Tayler
   p. cm.
ISBN 0 521 36431 0 (hardback). – ISBN 0 521 36710 7 (paperback)
1. Galaxies. 2. Galaxies – Evolution. I. Title.
OB857.T39 1993
523.1'12 – dc20 92-19071 CIP

ISBN 0 521 36431 0 hardback
ISBN 0 521 36710 7 paperback

Contents

Preface

This book is in effect a second edition of the book *Galaxies: Structure and Evolution* published by Wykeham Publications in 1978. When copies of the original edition were exhausted, the publishers were unwilling to reprint it. I am grateful to Dr Simon Mitton of the Cambridge University Press for agreeing to take the book over and for encouraging me to undertake the necessary task of revising the text.

The problem of the structure and evolution of galaxies is central to astronomy. On the one hand a galaxy is composed of stars, whose individual properties are known at least in broad outline. However, the process of star formation, which is crucial to the evolution of galaxies, is not at all well understood. On the other hand galaxies and clusters are the main constituents in the Universe and their properties provide important information about the origin and evolution of the Universe. In addition both the origin and present structure of galaxies are influenced by the possibility that the major form of matter in the Universe is not luminous stars but invisible weakly interacting particles.

In this book I discuss in general terms what is known both about the present structure of galaxies and about their past life history. Most of the detailed discussion refers to our own Galaxy. Although the subject is treated precisely where that is possible it will be apparent that, while the main ideas appear to be well-established, there are very considerable detailed uncertainties. When the previous version was published, it was just being realised that not all elliptical galaxies were spheroidal and that there was hidden matter in galaxies which could make their total masses much larger than was previously believed. Since then there has been much work in trying to understand the origin of galaxies and their distribution in the Universe, with the hidden matter playing an important rôle.

In a book of this size on such a complex topic it is impossible to give credit for every advance in the subject or to mention all of those people from whom I have learnt and from who I have borrowed material. The books which I have found particularly useful are mentioned in the Suggestions for Further Reading on page 204. My own interest in galaxies was greatly stimulated by discussions with Donald Lynden-Bell, Bernard Pagel and Martin Rees. I am grateful to Mrs Pauline Hinton for her careful typing of the manuscript.

I am happy once again to dedicate the book with respect and affection to Professor Jan H. Oort, who has made fundamental contributions to the study of galactic structure for more than sixty years. In particular 1992 sees the diamond jubilee of his first discussion of the Oort limit.

R. J. Tayler

February 1992

Note added in proof. Jan Oort died in November 1992. The book is now dedicated to his memory.

R. J. Tayler

Symbols

a, b	semi-major, -minor axis of ellipse
a, b	mass infall rate, dimensionless value
A, B	Oort's constants of galactic rotation
A	area of magnetic flux tube
B	magnitude of magnetic induction
$B_\perp$	transverse component of magnetic induction
$B_\nu(T)$	Planck function
d	distance of star from LSR, of closest approach of two stars
e	eccentricity of spheroid in galactic model
E, S0, S, SB, Irr, cD	types of galaxy
E_{cr}	energy density of cosmic rays
E_N	nuclear energy content of star
f	mass distribution function
$f(M)$	initial mass function
F	particle distribution function
$g_{\bar\omega}, g_z$	components of galactic gravitational field
H_0	Hubble's constant
$I, I_1–I_5$	integrals of motion of star
j	electric current density
l	galactic longitude, mean free path
L	galactic luminosity, scale of magnetic field
L_s	stellar luminosity
m	particle mass
M	mass, general or galactic
M_{HI}	mass of neutral hydrogen
M_P, M_{Sph}	point mass, spheroidal mass in galactic model
M_s	stellar mass
M_V	visual magnitude

n	number density
OBAFGKMRNS	spectral types of stars
p	yield of heavy elements
$P, P_{\bar{\omega}}$	period, general, epicyclic
P_{gas}, P_{cr}, P_{mag}	gas, cosmic ray, magnetic pressure
P_{rad}	radiated power radio source
q_0	deceleration parameter
r	spherical polar radius
r, R	radius, general
r_s	stellar radius
R, R_0	scale factor in Universe, present value
R_0	distance to galactic centre
R_{Sch}	Schwarzschild radius
S	mass of stars formed
t	time
t_H	Hubble time
t_{ms}	main sequence lifetime of star
T	temperature, total kinetic energy
T_e, T_s	effective, surface temperature of star
u, v, w	velocity components of star relative to Sun
$\bar{u}, \bar{v}, \bar{w}$	mean velocity components of stars
$u_\odot, v_\odot, w_\odot$	solar motion
v	velocity, general
v_{circ}	velocity of galactic rotation
v_{esc}	escape velocity
v_{gas}	velocity of gas
v_R, v_T	radial and tangential velocities of star
$v_{\bar{\omega}}, v_\phi, v_z$	velocity in cylindrical polar coordinates
$v_{\phi 0}$	velocity of Local Standard of Rest
x, y, z	cartesian coordinates
x, y	dimensionless value of star formation rate and of mass in stars
z	redshift
Z, Z_1	fractional mass of heavy elements, present value
α	semi-major axis of spheroid in galactic model, mass fraction locked up in dead stars
η	electrical resistivity
$\varkappa$	epicyclic frequency
λ	fractional mass converted into heavy elements and ejected from stars, wavelength
Λ	half thickness galactic disk
μ	proper motion of star
μ, μ_1	gas fraction in galaxy, present value
υ	frequency
ξ, η	displacement of star from circular orbit

$\tilde{\omega}, \phi, z$	cylindrical polar coordinates
ρ	density
ρ_0	critical density to close Universe
ρ_{gal}	smoothed out density of galactic matter
$\rho_{gas}, \rho_{cr}, \rho_{stars}$	gas, cosmic ray, star densities
σ	mass of gas in unit area of galactic disk
Σ	mass of stars formed
τ_c	collision time
τ_D	decay time of magnetic field
Φ	gravitational potential
ω	angular velocity
ω, ω_0	angular velocity of galactic rotation, at Sun
ω_s	spiral frequency
Ω	total gravitational potential energy
Ω_0	ratio of density of Universe to critical density
$\mathscr{E}, \mathscr{E}_{tot}$	energy of cosmic ray electron, total energy of system of electrons

Because of the large number of symbols required and of the desirability of conforming to standard usage, some symbols are used with more than one meaning. Which meaning is intended will be clear from the context.

Numerical values

Fundamental physical constants

a	radiation density constant	7.55×10^{-16} J m^3 K^{-4}
c	velocity of light	3.00×10^{8} m s^{-1}
e	charge on electron	1.60×10^{-19} C
G	gravitational constant	6.67×10^{-11} N m^2 kg^{-2}
h	Planck's constant	6.62×10^{-34} J s
k	Boltzmann's constant	1.38×10^{-23} J K^{-1}
m_e	mass of electron	9.11×10^{-31} kg
m_H	mass of hydrogen atom	1.67×10^{-27} kg
μ_0	permeability of free space	$4\pi \times 10^{-7}$ H m^{-1}

Astronomical quantities

$L_\odot$	luminosity of Sun	3.86×10^{26} W
$M_\odot$	mass of Sun	1.99×10^{30} kg

Approximate astronomical quantities

A	Oort's first constant	14 km s^{-1} kpc^{-1}
B	Oort's second constant	-12 km s^{-1} kpc^{-1}
H_0	Hubble's constant	50 km s^{-1} Mpc^{-1}
R_0	Distance to galactic centre	8.5 kpc
t_H	Hubble time	2.0×10^{10} yr
$v_{\phi 0}$	Circular velocity near Sun	220 km s^{-1}

Non SI Units

light year (unit of distance)	9.5×10^{15} m
parsec (unit of distance)	3.09×10^{16} m
year	3.16×10^{7} s

1

Introduction

The discovery of galaxies

In the medieval world picture the stars were regarded as points of light attached to a sphere, whose surface was a long way outside the Solar System but whose volume was thought to be very much smaller than the space which we now know the stars to occupy. Attempts had indeed been made to determine the distance to the stellar sphere based on the possibility that stars might appear to be in different directions when observed from two points on the Earth's surface (fig. 1). Although this method worked for the Sun and Moon and other objects in the Solar System, it failed for the stars, indicating that they were very distant. After the Scientific Revolution in the 16th and 17th centuries, culminating in Newton's explanation of the motion of the planets in terms of a universal law of gravitation, it was realised that the stars were probably also suns or equivalently that the Sun was but one star amongst many and that the *fixed stars* should, in fact, be moving through space and should be influenced by the same law of gravitation. This led to a renewed interest in trying to determine not only their positions but also their motions.

Initially it was thought that there was just one system of stars filling the Universe and this view persisted until the second decade of this century. Earlier than that it had become apparent that the stars are not uniformly distributed in space as, to a first approximation, the stars visible to the naked eye appear to be. At the end of the 18th century William Herschel made the assumption that all stars have the same absolute light output as the Sun and concluded that the visible system of stars had an edge. Although his assumption about stellar luminosities was incorrect, his conclusion was not. By the beginning of the 20th century it was known that the system of stars which we now refer to as our *Galaxy* was highly flattened, most of the stars falling close to a plane which is defined in the sky by the faint band of light known as the *Milky Way*. At the same time it was known that distributed around the sky there were large numbers of fuzzy luminous objects which were clearly not

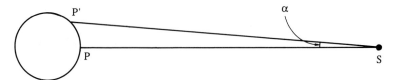

Figure 1.   An attempt to measure the distance to a star. If stars were sufficiently close to the Earth, it would be possible to measure the angle α between the direction of a star as seen from two points P, P′ on the Earth's surface. With α and PP′ known, SP could be calculated.

individual stars and which had been given the name *nebulae*. Herschel and his son John were pioneers in cataloguing nebulae and the largest list was the New General Catalogue (NGC) of Dreyer.

Some of these nebulae were subsequently shown to be true nebulae or clouds of gas, which was what their name was meant to indicate, and they are situated in the same region of space as the stars. In contrast, one group of nebulae in particular, known as the *spiral nebulae*, were in time discovered to be composed of faint stars. There was then considerable argument early in the 20th century about whether the spiral nebulae were situated in the outer regions of our Galaxy or whether they were independent stellar systems, possibly of similar status to our Galaxy. If their stars were in fact similar to nearby stars, it was obvious that they must be distant and independent stellar systems, since their stars appeared so faint. This was eventually shown to be true in the early 1920s and these nebulae were sub-sequently given the name *galaxies*.†

It is now known that the observable Universe contains thousands of millions of galaxies and this number may be a considerable underestimate because of the difficulty of detecting small faint galaxies and any galaxy which has a low surface brightness. The Galaxy is a large galaxy but certainly not one of the very largest. It contains well over 10^{11} stars but the largest galaxies probably contain 10^{12} or 10^{13} stars. Obviously all of these stars have not been counted, although with modern telescopes and instruments which count stars automatically, many millions of stars can be counted in our own Galaxy. Very detailed studies have been made of limited regions of the Galaxy. Although galaxies contain other matter in the form of *interstellar gas* as well as stars, it appears that most of the visible mass in galaxies is in the form of stars, at least at the present time. We can therefore, to a first approximation, regard galaxies simply as systems of stars. It will however become clear later that galaxies also contain much invisible matter and that this matter may be neither stars nor gas. It is generally believed that much of it is composed of elementary particles. These particles, both inside and outside galaxies, may be the main form of mass in the Universe and may play a key rôle in the formation and structure of galaxies. Both stars and galaxies are held together by the force of gravitation and one of the most important questions in astronomy is *why matter in*

† Our Galaxy will always be spelt with a capital G and will be called *the Galaxy* or *our Galaxy*. The name galaxy is derived from the Greek name for the Milky Way.

the Universe is arranged in objects of galactic mass which have inside them sub-units which are objects of stellar mass. Although I shall not be able to answer this question in this book I shall try to present the evidence upon which an answer must be based.

Much of the detailed knowledge which we possess about stars has been obtained because we are fortunate enough to be close to one particular star – the Sun. In the same way we might hope to learn much about galaxies by a study of the Galaxy. There is, however, one very important distinction; whereas we are *near* to the Sun, we are *inside* the Galaxy. It proves quite difficult to discover the structure of an object from the inside. As an example, for a long time it was believed that the Sun was very near to the centre of the Galaxy because the apparently bright stars are approximately symmetrically placed around the Sun. It was subsequently realised that, as will be described further below, the interstellar gas in the Galaxy absorbs starlight and behaves as an interstellar fog, which gives a totally wrong impression about distances inside the Galaxy.

Differences between stars and galaxies

Although both stars and galaxies are objects which are held together by the attractive force of gravitation, they differ in many important respects both qualitative and quantitative. One simple observational fact is that, whereas the majority of stars are spherical or depart only slightly from spherical shape, galaxies exist in many shapes from essentially spherical to ones which are highly flattened; some of the latter are spheroidal but others have a much less symmetrical structure. Many highly flattened galaxies are observed to rotate rapidly. The great variety in galactic shape indicates that the classification of galaxies may be much more complicated than the classification of stars. Those properties of the classification of stars which are required in this book will be introduced as they are needed; the classification of galaxies is discussed in Chapter 3.

Another very important difference between stars and galaxies is that there is at present no clear evidence that there exist galaxies of significantly different ages; they may almost all have been formed between 10^{10} and 2×10^{10} years ago, with the actual spread of ages being much less than either of these figures. We shall, however, also see that the formation process of a galaxy may be very prolonged so that there is not a sharp distinction between formation and evolution. This is totally different from the case of stars in the Galaxy, where we know that some are essentially as old as the Galaxy while others are no more than a few million years old and star formation is certainly still continuing in the Galaxy today. It has proved possible to study stellar evolution not by observing the changes with time in the properties of an individual star but by investigating the properties of stars of similar masses but of different ages in our neighbourhood of the Galaxy. As all the galaxies near to us are about equally old, it appears that our only hope of a direct study of galactic evolution is to look at the most distant galaxies in the Universe and to compare their properties with those of nearby galaxies. Because light has taken a considerable time to reach us from them, we see them as they were in the remote past and, if the average properties of galaxies change as they age, we might hope to obtain some information about this change. At present the information that has been obtained is not very clear. One reason is that the distant galaxies appear very faint and are difficult to study. Another reason is that the assumption that galactic properties do *not* change significantly with *age* tends itself to be used in deriving the *distance scale of the Universe,* as will be discussed later in this chapter. What is clear is that the timescale of significant galactic evolution appears to be comparable with the believed *age of the Universe.*

The evolution of galaxies

There are two independent reasons which suggest that the Galaxy (as well as most other galaxies) may have been much more luminous early in its life history than it is now. If the Galaxy was formed by condensation from intergalactic gas, as is generally thought likely, it must initially have had a much greater total energy than it at present possesses. Thus a collapsing cloud has sufficient energy to expand again to its initial size assuming no dissipation of energy occurs and, if it is to settle down as a relatively compact object compared with its initial size, a considerable amount of energy must be lost from the system. However the energy is lost, it must be lost during the initial collapse phase which may last only a very short time compared to the total life of the Galaxy. This problem of galactic formation will be discussed further in Chapters 7 & 8. The second reason, which relates specifically to our own Galaxy but which almost certainly applies to other galaxies as well, is concerned with its chemical composition and this will be discussed in Chapter 7. If, as seems plausible, the original chemical composition of the Universe was a mixture of hydrogen and helium and if the heavier elements have all been produced by nuclear reactions in stars or more massive objects in the galactic lifetime, the observed variation of stellar chemical composition with stellar age suggests that there must have been a rapid burst of production of heavy elements on a galactic scale early in galactic history. This also suggests a very high early galactic luminosity. We shall, however, see in Chapter 7 that there are some suggestions that these conclusions may not be correct and that the initial collapse phase must be much longer than was originally believed.

Determination of the structure of the Galaxy

I now consider in slightly more detail how the present structure of the Universe has been deduced. As I have already mentioned, the discovery of Newton's law of universal gravitation and its application to the motion of bodies in the solar system led to the realisation that stars should be both distributed at different distances through space and moving. This stimulated attempts both to determine the distances of stars and to study their motions. It was not very long before Halley discovered the motion of stars across the sky through the background of (presumably) more distant stars (fig. 2). This effect, which was given the name *proper motion*, was discovered by comparing the positions of stars in the sky

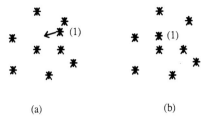

(a)   (b)

Figure 2.   Proper motion. In (a) the star marked (1) is moving in the direction of the arrow. As a result, at a later time in (b), its position relative to more distant or more slowly moving stars has changed. Its angular displacement is called its proper motion.

according to the observations of the 17th century with those made by the Greeks many centuries earlier. It is now realised that there was an error in Halley's procedure which vitiated his results but reliable proper motions were subsequently measured. The discovery of proper motions will become very much easier as observations accumulate over a very long period. There is some advantage if successive observations are made with the same telescope, so that photographic plates can be compared directly, and the need for such observations has ensured that some very old telescopes have remained in use long after they would otherwise have been scrapped. However, very accurate modern measuring techniques are making it easier than it was to compare results obtained with different telescopes. Both proper motions and parallaxes (defined below) are being measured by the HIPPARCOS satellite launched in 1990. Unfortunately the satellite did not get into its planned orbit and as a result its lifetime will be reduced. Nevertheless it appears that very valuable results are being obtained.

It took very much longer to measure the distance to any stars. Both Newton and Herschel estimated the distances to stars by assuming that they were all really just as bright as the Sun, so that their apparent brightness was a measure of their distance. Because the Sun is a fairly average type of star, these estimates turn out to be very good for many stars although they are very bad indeed for some very bright or faint stars. The estimates were good enough to show that stars are extremely distant in comparison with the size of the solar system. However, the first direct measures of stellar distance were not made until three astronomers obtained distances to different stars in 1838/39 by the method of *stellar parallax*† (fig. 3); the parallax is the angle subtended at the star by the radius of the Earth's orbit around the Sun. For all stars that have so far been discovered it is less than 1″.

Once the distance of a star is known as well as its proper motion, the angular velocity across the sky can be converted into a *tangential velocity,* usually expressed in km s^{-1}. Finally the development of the science of spectroscopy in the second half of the 19th century led to the measurement of stellar *radial velocities* by use of the Doppler effect (fig. 4). By the early years of the 20th century it had

† The three stars were 61 Cygni measured by F.W.Bessel, α Lyrae (Vega) by F.G.W. Struve and α Centauri by T. Henderson.

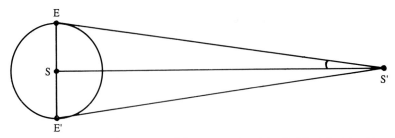

Figure 3.   Parallax. If a star S' is observed from opposite sides E,E' of the Earth's orbit around the Sun S, the angle ES'S is called its parallax. If the angle is measured, the distance to the star can be determined.

become apparent that the stars in the neighbourhood of the Sun were typically a few light years† apart and that they had velocities relative to the Sun of a few tens or in some cases a few hundreds of kilometres per second.

By that time it had become clear that the Sun was one member of a large dynamical system of stars and it was believed that it was very close to the centre of the system. The reason for that belief is as follows. I have already remarked that the stars visible to the naked eye are approximately symmetrically placed about the Sun in space. In addition the Milky Way is a belt of fainter and more distant stars. This led to the picture of the Galaxy shown in fig. 5. A central, approximately spherical, distribution of stars is surrounded by a belt of more distant stars in the form of a torus; that is, shaped like a car tyre inner tube. This view of the Galaxy (or the Universe as it was then supposed to be) was put forward by Eddington in his book *Stellar Movements and the Structure of the Universe* (1912). Only ten years later a very different view of the Galaxy was gaining acceptance; this is shown in fig. 6. This shows an edge-on view of the Galaxy with the Sun very far from the central nuclear bulge. How could such a dramatic change of view have come about in such a short time?

The distribution of globular clusters

One clue to the change of view is given by the *globular star clusters* which are shown in fig. 6. These are compact groupings of stars which contain perhaps 10^5 to 10^6 stars each. The American astronomer Shapley pointed out that they do not appear in all directions in the sky, as is obvious from the position of the Sun and Solar System in fig. 6, but that they do appear to be part of the Galaxy. He suggested that the globular clusters form an approximately spherical system centred on the centre of our Galaxy. He then found that he could make sense of the observations only if the Sun were a large distance from the centre of the Galaxy (approximately 15 kpc) comparable with the radius of the system of globular clusters. This is shown in fig. 6.

† Throughout this book we shall use one or other of two measurements of distance. The *light year,* the distance travelled by light in one year, is 9×10^{15} m and the *parsec,* the distance at which a stellar parallax is one second, is 3×10^{16} m. Inside galaxies the useful unit is 10^3 pc(kpc), between galaxies 10^6 pc(Mpc) and for the whole observable Universe 10^9 pc(Gpc).

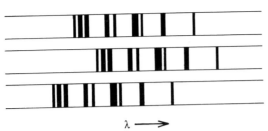

Figure 4.   The Doppler shift of stellar spectra. The three spectra represent respectively that of a star at rest relative to the Sun, one receding and one approaching the Sun. λ is wavelength.

Figure 5.   Eddington's (1912) picture of the Galaxy. The Sun is marked **x**.

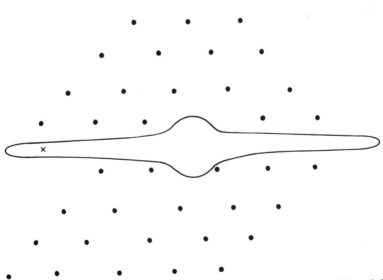

Figure 6.   A schematic view of the Galaxy from the side showing the thin disk and the central nuclear bulge. The filled circles are globular star clusters and the Sun is marked **x**.

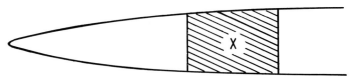

Figure 7.   Effect of interstellar obscuration. Because of the interstellar obscuration it is only easy to see disk stars in the small hatched region around the Sun (**x**).

Interstellar gas and dust

Although this explains the observed distribution of globular clusters, it is now necessary to understand why this eccentric position of the Sun is not immediately apparent in our observations of stars. In particular, why is the sky not overwhelmingly more bright in the direction of the galactic centre than else-where? The answer to this question soon became apparent with confirmation of the existence of significant amounts of *interstellar gas and dust* which had been suspected much earlier. The way in which this gas and dust have been discovered will be described in Chapter 2; for the present it is sufficient to say that the interstellar matter absorbs and scatters starlight. The interstellar dust produces an interstellar fog, which makes it very difficult for us to see distant stars, so that the stars which we see with the naked eye or a small telescope are essentially only those near to the Sun (fig. 7). The gas and dust lie in the thin disk of the Galaxy, which also contains most of the stars. As many of the globular clusters are well out of the disk, as is indicated in fig. 6, there is not much interference with the study of the system of globular clusters. There is, however, *some* absorption of starlight in the direction of the globular clusters and, partly as a result, the present best estimate of the distance to the galactic centre (8.5 kpc) is significantly less than the value obtained by Shapley.

External galaxies

At the same time that the outline structure of the Galaxy was being resolved, there was dispute about whether the Galaxy was the whole Universe or whether there were other galaxies. As I have already mentioned, the dispute concerned objects known as nebulae. It had been known for a very long time that in addition to stars the Universe contained nebulous looking objects. The nebulae were studied by astronomers such as William Herschel and were catalogued by the French astronomer Messier towards the end of the 18th century. A much more complete list was the *New General Catalogue* produced by Dreyer almost a hundred years later. Even today many galaxies are known by a number given in one or other of these catalogues. The Andromeda galaxy is M31 and one of its satellite galaxies is NGC205. The nebulous objects were ultimately shown to fall into essentially four classes; star clusters (such as globular clusters), true nebulae (gas clouds), planetary nebulae (a cloud of gas surrounding a star, which has ejected it) and the group which we now know to be galaxies. The principal

Figure 8.   The zone of avoidance. Because of absorption by interstellar matter near the galactic plane, very few galaxies can be seen from the Sun (**x**) in directions between the two dashed lines.

examples of this group were known as *spiral nebulae* because of their spiral appearance, first discovered by the Earl of Rosse. Already, long before the spiral structure was known, an Englishman, Thomas Wright, and a German, Immanuel Kant, had suggested that some of the nebulae might be *island universes*.

There were two main reasons why the discovery that the spiral nebulae were independent galaxies was delayed. The first was that they are so distant that they could not be shown to be systems of stars until very large optical telescopes were available. This observational situation was transformed when the 100 inch Mount Wilson telescope came into operation in 1919. The second reason was concerned with the arrangement of the nebulae in space. The nebulae, unlike the bright stars, did not appear to be completely uniformly distributed around the sky. Instead they appeared to be largely absent in the direction of the plane of our Galaxy, giving rise to what was called the *zone of avoidance* (fig. 8). Very few nebulae could be seen in directions in and close to the plane of the Milky Way. Because through this observation the system of spiral nebulae appeared to be related to the structure of our Galaxy, there were very strong arguments that they must be part of the Galaxy or at most nearby satellites of it. These arguments were realised to be invalid when the discovery of the absorption of starlight by interstellar matter indicated that galaxies in the direction of the zone of avoidance could not be seen even if they were there. At about the same time, it became possible to see, on photographs taken with the 100 inch telescope, that the nearby spiral galaxies were composed of stars. If these stars were assumed to be similar to stars in our Galaxy, it was possible to estimate that nearby galaxies were typically several million light years away and consequently well outside our Galaxy. Thus, the idea of a Universe of galaxies became established.

The expanding Universe

Possibly the most compelling evidence for the independence of the external galaxies was obtained in the middle 1920s largely through the work of Hubble and his collaborators. I have already shown that the radial velocities of stars in the Galaxy can be measured by means of the Doppler shift of spectral lines. Hubble measured the Doppler shifts for many galaxies instead of stars. In the case of nearby galaxies, the Doppler shift could be either a redshift or a blueshift indicating that there are random motions in the system of galaxies similar to the random velocities of stars in the solar neighbourhood. In contrast the spectral lines of all of the more distant galaxies were shifted to the red. If the redshift is interpreted as being due to the Doppler effect, this means that all of the distant galaxies have a relative velocity directed away from us. This observation led to the idea of the *expansion of the Universe*. A discussion of the expanding Universe is the subject matter of *cosmology*, which I shall not study in detail in this book. It is not, however, possible to interpret the observations of distant galaxies or to discuss the structure and evolution of galaxies without taking some notice of cosmological problems, as I shall explain in both Chapters 3 and 8. Other suggestions have from time to time been advanced to explain the redshift. These include the *tired light* hypothesis that light loses energy and is redshifted simply by travelling large distances. Neither this nor any other suggestion appears to be as well-founded as the Doppler explanation and I shall assume without further discussion that the Universe *is* expanding. This assumption will only significantly affect the contents of Chapter 8.

The general picture of the Universe which was established in the 1920s is still generally accepted today. Many new types of object have been discovered inside galaxies, particularly by the use of observational techniques in new branches of astronomy such as *radio astronomy, infrared astronomy, ultraviolet astronomy* and *X-ray astronomy*. In fact the experience of the present century has been that, each time a new technique has become available, a whole class of exciting new objects has been discovered. Most of the galaxies in our neighbourhood have fairly regular shapes but recently many so-called peculiar galaxies have been discovered. These include galaxies which are apparently undergoing an explosion and others which have been distorted by the gravitational fields of their neighbours. In addition the *quasars* have been discovered. They have the largest redshifts of any objects known, they are extremely luminous and there is not as yet a full understanding of the source of the luminosity. The properties of peculiar galaxies and quasars will be described briefly in Chapter 3.

The Hubble constant

When Hubble discovered the expansion of the Universe, he also provided an estimate of the distance to the galaxies which he observed. It is now believed that his estimates are almost a factor of ten too low. I shall shortly explain how this has happened. This estimate can be summed up in one parameter, the *Hubble constant*. Hubble showed that the velocity of recession of distant galaxies appeared to be related to their distance from us by a law of the form

$$v = H_0 r, \tag{1.1}$$

where H_0 is known as Hubble's constant. With v usually measured in km s^{-1} and r in Mpc, the units of H_0 are km s^{-1} Mpc^{-1}; clearly H_0^{-1} has the dimensions of time and it (t_H) is in some sense *the age of the Universe*. I shall make this relation more precise in Chapter 8. An early value of H_0 obtained by Hubble was

$$H_0 \approx 550 \text{ km s}^{-1} \text{ Mpc}^{-1}, \tag{1.2}$$

whereas the most favoured value† today is probably

$$H_0 \approx 50 \text{ km s}^{-1} \text{ Mpc}^{-1}. \tag{1.3}$$

The corresponding Hubble times are

$$t_{\mathrm{H}} \approx 1.8 \times 10^9 \text{ yr} \tag{1.4}$$

and

$$t_{\mathrm{H}} \approx 2 \times 10^{10} \text{ yr}. \tag{1.5}$$

The distance scale of the Universe – standard candles

Throughout this book I shall be assuming that the distances to galaxies are known to a reasonable accuracy. For this reason, at least, it is desirable for an account to be given of how distances in the Universe are estimated and of how such large variations in the believed value of H_0 have arisen. It is not possible in a few pages to give a complete account of the establishment of the cosmic distance scale. The most that can be done is to describe the steps involved. The first point which must be made is that the direct measurement of distance by the use of stellar parallax does not carry us very far. The nearest star has a parallax of less than 1″. It seems hardly possible that such an angular displacement of one star relative to the background of more distant stars can be measured accurately and in particular that the first parallaxes were obtained before the use of photography in astronomy. In fact, much smaller parallaxes can be measured but the smallest that are even reasonably reliable are about (1/50)″. This means that it is possible to obtain the distances of stars out to about 50 parsecs. When it is realised that the centre of our own Galaxy is about 8.5 kpc distant, it can be seen that direct measurements of distance do not take us very far. Beyond 50 pc we must rely on indirect methods. All of these depend on the use of a *standard candle*. I define a standard candle to be a type of object which has a property, or properties, which is really well known. If I can then recognise it at large distances, I can combine an observation of its apparent brightness or apparent size (if brightness or size is the property which is well known) with a knowledge of its true brightness or size to give its distance. The main problem then is the discovery and calibration of such standard candles. The changes in the estimated distances to galaxies have arisen largely because of mistakes in the identification of standard candles.

Cepheid variable stars

The first step is the establishment of distances inside the Galaxy and to the nearest galaxies. Two main techniques are used here; *Cepheid variables* and *cluster main sequence fitting*. Cepheid variables are a class of star whose brightness

---

† There is still considerable controversy about the value of the Hubble constant with some workers favouring values as high as 100 km s^{-1} Mpc^{-1} and with others obtaining values as low as 40 km s^{-1} Mpc^{-1}. Later in this book when I quote distances or other quantities whose values depend on the value chosen for H_0, I shall use 50 km s^{-1} Mpc^{-1} unless I specifically display a dependence on the value of H_0.

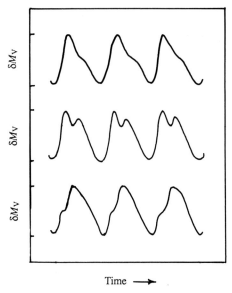

Figure 9.   Light curves of Cepheid variables. The characteristic shape of the light curve changes with the period of the oscillation. The top curve is that of the Cepheid of shortest period (M_V is proportional to minus the logarithm of the luminosity L_s).

varies periodically as shown in fig. 9. Although the light curve of a variable is not completely smooth, it is well-defined for any one variable and there is a general relationship between the shape of the light curve and the period of light variation. What is more important is that from a study of Cepheid variables in a satellite galaxy of our own, the *Small Magellanic Cloud*, it was shown that there is a very close relationship between the period and luminosity (total light output) of a Cepheid. This relationship could be established for the Cepheids in the Magellanic Cloud because, although their distances from us are not known, the differences between their distances are very much smaller than any one distance. Thus they can all be regarded as being at the same distance and the observed relation between period and apparent luminosity can equally be regarded as a relation between period and absolute luminosity.

We now need to calibrate this relation. We need to know the absolute luminosities of a number of Cepheids. We will know this if we can determine the distances to a small number of Cepheids in our Galaxy. Unfortunately there is no Cepheid which is near enough for its parallax to be measured. For a long time this posed a problem. The distance to nearby galaxies could be determined in terms of the distance to the Small Magellanic Cloud but that was not well known. Then a small number of Cepheids were found to be members of star clusters in our Galaxy known as *galactic* or *open clusters*. These star clusters are found in the disk of the Galaxy. They contain fewer stars than the globular clusters, which have already been mentioned, but they are also close groupings of stars which are gravitationally bound systems. If the distances of the galactic clusters can be obtained, we

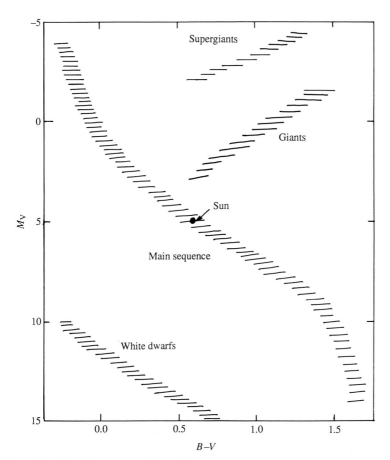

Figure 10. The Hertzsprung-Russell diagram for nearby stars. The visual magnitude M_V is plotted against colour index $B - V$ and most stars fall in four well-defined groups. (M_V is proportional to $-\log L_s$ and $B - V$ is related to the logarithm of surface temperature).

know the absolute luminosities of some Cepheids. As we also know their periods, the period/luminosity relation can then be calibrated. To obtain the distances to galactic (or globular) clusters we must use some observations of the properties of the nearest stars whose distances *are* known.

Cluster main sequence fitting

If I plot the luminosities of the stars of known distance against their surface temperatures, I obtain a diagram of the form of fig. 10, the *Hertzsprung–Russell (H–R) diagram*. In this diagram about 90 per cent of the known nearby stars lie in the region known as the *main sequence*. These are believed to be stars which, like the Sun, are at present radiating energy which has been released by nuclear processes converting hydrogen into helium. If I next study the stars in a

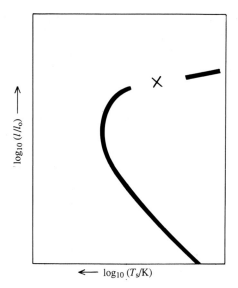

Figure 11.   The Hertzsprung-Russell diagram for a galactic star cluster. The logarithm of apparent luminosity is plotted against logarithm of surface temperature. **x** marks the *Hertzsprung gap* in which very few stars are found.

(galactic) cluster, I can plot *apparent* luminosity against surface temperature and I might obtain a diagram of the form shown in fig. 11. For our present purposes only one feature of this diagram is important. This is that we again have a group of stars forming what looks like a main sequence, although the top of the sequence is absent. If I suppose that the absolute luminosity of stars on the cluster main sequence is the same as the luminosity of stars of similar surface temperature on the main sequence of nearby stars, I can estimate the distance of the cluster. This is known as main sequence fitting. It is believed to give reasonably reliable distances to both galactic and globular clusters in the Galaxy. As it also gives the distance of about fifteen Cepheid variables, the period-luminosity relation can be calibrated.

Further steps in the distance scale

With the Cepheid relationship established, we can now obtain distances to those galaxies in which Cepheids can be observed. This takes us out to about 4 Mpc. Beyond that distance we need further standard candles. We first choose constituents of galaxies which are more luminous than Cepheids. These include:

    (a) The brightest visible stars in a galaxy,
    (b) Globular clusters,
    (c) HII regions,
    (d) Novae,
    (e) Supernovae,
    (f) Planetary nebulae.

Of these (a) needs no definition and I have already introduced (b). *HII regions* are luminous clouds of ionised gas (mainly hydrogen, hence the name, HII being the spectroscopic designation for ionised hydrogen) which are brighter in the visible region of the spectrum than the brightest stars and they can therefore be seen in more distant galaxies. They also have an extended image rather than appearing as a point source so that their apparent diameters can be measured. They are in fact ionised by the most luminous stars, which radiate principally in the ultraviolet, and they effectively hide the stars from view. *Novae* are stars which suddenly become much brighter than they were before and subsequently fade in brightness, returning approximately to their original state. We can study their maximum brightness in our own and neighbouring galaxies and it appears that the brightest novae may be reasonably standard candles. Supernovae are even more violent and luminous stellar explosions which destroy a star. There are two main types of supernovae and the properties of Type I supernovae are quite homogeneous. They provide a very promising way of extending the distance scale. Planetary nebulae are gas clouds which have been expelled by their central stars and studies of nearby galaxies show a similarity in the distribution of luminosities of their planetary nebulae.

If all galaxies were the same shape and size it might be reasonable to suppose that the luminosity of their brightest stars, the sizes of their HII regions, and the luminosities of their novae and planetary nebulae would be the same. It is, however, obvious from a study of galaxies whose distances are known from the Cepheid method that there is a wide variety of galactic properties. I will defer a detailed discussion of galactic classification until Chapter 3, but remark now that the two main types of galaxy are *spiral galaxies* and *elliptical galaxies*. Photographs of nearby spiral and elliptical galaxies are shown in figs. 12 and 13. Spiral galaxies are further classified according to the tightness of their spiral structure and ellipticals according to their degree of apparent departure from spherical shape. In using standard candles of the type mentioned above, it is necessary to compare galaxies of similar type. Very poor results would be obtained if the contents of a small spiral galaxy were compared with those of a giant elliptical galaxy. The standard candles which have been mainly used are brightest stars (or more reliably the average of the three brightest blue stars which itself varies with galactic type or the brightest red stars) and the HII regions, and the HII regions are believed to give quite good distances out to about 25 Mpc. One of the reasons why Hubble's original estimates of distance in the Universe were invalid is that he wrongly thought that HII regions in some galaxies were stars. As a result he regarded them as objects whose absolute luminosities were those of stars rather than the higher luminosity of a bright HII region. This meant that he underestimated the distance to the galaxies.

At greater distances than those for which HII regions or possibly globular clusters can be used as standard candles we are forced to use the properties of entire galaxies. We can classify galaxies into different types; as already mentioned this classification of galaxies will be discussed in Chapter 3. It can then be assumed that galaxies of a particular type have the same size and/or luminosity or more

Figure 12.   A spiral galaxy M31, the great spiral in Andromeda. Also shown are two satellite elliptical galaxies. (Photograph from the Hale Observatories.)

precisely that the average properties of a type of galaxy do not vary with distance from us. As the classification of galaxies is based on structural details, shape, presence and type of spiral arms, etc., which can be recognised out to very great distances, in principle this method can carry us very much further. Beyond that we can recognise that many galaxies are found to be members of clusters of galaxies and we can try to use statistical properties of clusters of galaxies to provide us with standard candles for the later parts of the cosmic distance scale.

Figure 13.   A nearby dwarf elliptical galaxy (NGC 147) in Andromeda, showing resolution into stars. (Photograph from the Hale Observatories.)

I should explain how the large uncertainty in the value of H_0 arises. It is not difficult to see that, if galaxies are standard candles, Hubble's velocity/distance relation is valid in terms of the relative distances of the galaxies. To obtain a value for H_0 we need the actual distance to a relatively nearby standard candle. To obtain this the secondary distance indicators described above must be used to obtain distances to galaxies whose random velocities, which may be towards or away from our Galaxy, are small compared with their recession velocities due to the expansion of the Universe. This involves distances of a few times 10 Mpc and it is this procedure which leads to the large differences between different estimates of H_0. We shall see later that the matter is further complicated because the Galaxy itself has a random velocity which causes the expansion about it to be not quite symmetrical.

Hidden matter

A property of galaxies which has only become fully apparent since the 1970s is that many of them contain much more matter than can be observed. The motions of stars in galaxies and of galaxies in clusters of galaxies are controlled by the force of gravitation. If the rotation of spiral galaxies is studied it is found that the mass necessary to explain the rotation, particularly of the outer regions, is significantly greater than the visible matter. The same result is true of the motions of galaxies in clusters of galaxies, where the *hidden mass* may be even more dominant. There is now a general belief that it is not ordinary matter, the chemical elements, which make up the solar system, stars and gas clouds, but that it is composed of what are known as weakly interacting massive particles. This will be discussed further in Chapter 8. It is however possible to discuss many of the properties of galaxies without mentioning hidden matter and I shall do this.

Outline of the contents of the present book

The uncertainties which I have just discussed will not trouble us in most of this book, where I shall be concerned mainly with our Galaxy and with other galaxies of a variety of types but of similar age to our own Galaxy. Henceforth I shall regard galactic distances as known, except when what I am discussing may itself have some influence on the determination of distances. The next two chapters will be devoted to a discussion of the observed properties of galaxies. Chapter 2 is concerned with our Galaxy and Chapter 3 with other galaxies. This division is to a large extent appropriate as the techniques involved in studying our own Galaxy tend to be different from those involved in other galaxies. Although the two chapters are supposed to be concerned with observations, we shall find that there are very few pure observations and that most of them must be combined with some theoretical ideas before they can be interpreted properly. In fact, we shall find that there is not really a fully logical order for the material in this book. Most aspects of the subject interact with other aspects. We have already seen this in that I cannot give a thorough discussion of the cosmic distance scale without simultaneously discussing the properties of our Galaxy and other galaxies.

Following the two chapters which are primarily concerned with observations, there are five chapters in which theoretical considerations are more prominent. Chapter 4 stands apart somewhat from the other material in this book, although some of its results are used in later chapters. In it a galaxy is considered primarily as a galaxy of stars, which can be regarded as the molecules in a gas, so that the galaxy can be described by statistical methods well known from the kinetic theory of gases. In Chapter 5 I discuss how the masses of galaxies can be determined. In a sense the mass of a galaxy can be considered to be an observational parameter so that this discussion could have been placed in Chapters 2 and 3. However, the theoretical contribution in getting from an observation to a mass is very large and it seems sensible to devote a whole chapter to this topic. In Chapter 6 I discuss the rôle of constituents of galaxies other than stars (gas and dust, cosmic rays and magnetic field) in the overall structure of galaxies. Chapter 7 on the chemical evolution of galaxies is concerned with the interstellar gas in a different way. It discusses how the chemical composition of the interstellar medium and hence of newly formed stars changes as a galaxy evolves, because heavy elements are produced from light elements through nuclear reactions in stars and some of the processed material is expelled from stars into the interstellar medium. In Chapter 8 I reach the most uncertain part of the book which is a discussion of the formation and early evolution of galaxies and of related cosmological problems. Finally an attempt is made to sum up what is known and what remains to be discovered in Chapter 9.

The subject of the structure and evolution of galaxies is one which is under very active theoretical and observational study at the present time. Although it is a time of rapid progress it is also true that many of the new discoveries are disturbing what were previously believed to be established facts. I have tried in the remainder of this book to stress the fundamental ideas of the subject and to avoid sounding certain when certainty is not at present appropriate. This book will have succeeded if it has convinced its readers that the subject is exciting and that it is therefore worthwhile trying to keep in touch with developments in the field. At present it seems likely that it will be decades rather than years before a fairly definitive book can be written.

2

Observations of the Galaxy

Introduction

This chapter contains a description of the properties of the Galaxy. It is concerned mainly with obervations, although as I have mentioned on page 18 many of the observations require a considerable amount of interpretation before they are very useful. The Galaxy is primarily a system of stars and I start this chapter by summarising some of the properties of stars of different types. As I shall explain later, there is some considerable uncertainty about the total mass of the Galaxy and about the masses of its individual components. In particular we shall learn that much of the mass of our own Galaxy and other galaxies may be invisible. Although this *hidden matter* might be very low luminosity stars or dead stellar remnants, there is a general belief that it is composed of weakly interacting elementary particles. At present I shall concentrate attention on the visible components. For them it may not be too far wrong to suppose that 95 per cent of the mass is stellar (including dead remnants) and about 5 per cent is in the form of interstellar *gas* and *dust*. In addition the Galaxy contains *cosmic rays*, very high energy charged particles, which contribute very little to the total mass but whose total energy is very important in discussions of the structure of the interstellar medium, as we shall see in Chapter 6. Finally, the galactic gas is pervaded by a *magnetic field* which is important in its own right and which has contributed to the growth of our knowledge of the Galaxy. One property of the magnetic field is that it constrains the motions of the cosmic rays in such a way that they spend much longer in the Galaxy than they would otherwise. I shall now discuss all of the components of the Galaxy in turn, although I shall reserve some detailed discussion of the non-stellar components until Chapter 6. In the first instance I consider the system of stars.

Properties of stars

I have already described how we can obtain information about the distances, proper motions and radial velocities of the nearby stars. As I have explained, these observations are to a large extent restricted to stars in our immediate neighbourhood, particularly because of the difficulty in measuring parallaxes. If we are to study the properties of individual stars even in a rather more extended region, we shall need other ways of estimating distances to supplement parallaxes and the other techniques, use of Cepheid variables, cluster main sequence fitting, discussed in Chapter 1. Cepheid variables are a particular and rare type of star and cluster main sequence fitting only gives distances to star clusters and not to individual stars. Before I discuss another method of estimating distances, I must summarise some properties of stars which will be important in this book.

I have already shown the Hertzsprung–Russell diagram for nearby stars in fig. 10 on page 13 and I have further explained how comparison with the H–R diagram of a star cluster (fig. 11) leads to the technique of main sequence fitting for finding the distances to star clusters. Most stars *are* main sequence stars so that in most cases a reasonable estimate of the distance to an individual star can be obtained by assuming that its luminosity is equal to that of a main sequence star with the same surface temperature. This method will give a completely wrong answer for the small fraction of stars that are not main sequence stars, i.e. giants, supergiants or white dwarfs. Fortunately there are ways of deciding whether or not a star is in one of these categories which do not require a knowledge of the stellar distance. These methods depend on a study of the detailed properties of the light emitted by a star: its *spectrum*.

Spectral types

Stars have been classified into *spectral types* with the classification depending on which spectral lines are most prominent in the spectrum of a star. When the classification was introduced with the classes being labelled A, B, C, . . ., the order of the classes was somewhat arbitrary and it was believed that stars of different spectral type had distinctly different chemical compositions corresponding to those elements whose spectral lines were strongest. Subsequently, it was shown that in most cases the important factor determining which spectral lines were prominent and hence determining spectral type was the surface temperature of the star and not its chemical composition. The surface temperature determines whether an element is in the correct state of ionisation and excitation for its spectral lines to be excited. Some of the original spectral classes were then shown to be not really significant and the remainder can be arranged to give a sequence of decreasing surface temperature

O B A F G K M R N S.†

The important spectral features of each class in the visible region of the spectrum and the corresponding surface temperatures are shown in Table 1. Note that the

---

† This order is most readily remembered by use of the mnemonic Oh be a fine girl kiss me right now, sweetheart. If the mnemonic is now thought sexist, girl can be replaced by guy.

Table 1. *Main features in the spectrum and approximate surface temperatures (in K) of stars of different spectral types.*

O	Ionised helium and metals, weak hydrogen	5×10^4
B	Neutral helium, ionised metals, hydrogen stronger	2×10^4
A	Balmer lines of hydrogen dominate, singly ionised metals	10^4
F	Hydrogen weaker, neutral and singly ionised metals	7.5×10^3
G	Singly ionised calcium most prominent, hydrogen weaker, neutral metals	6×10^3
K	Neutral metals, molecular bands appearing	5×10^3
M	Titanium oxide dominant, neutral metals	3.5×10^3
R,N	CN, CH, neutral metals	3×10^3
S	Zirconium oxide, neutral metals	3×10^3

straightforward temperature sequence ends at class K and that small differences of chemical composition do affect the later spectral types, where molecular bands are important spectral features. Stars with classification O, B, A are referred to as stars with *early spectral type*; F and G as *intermediate spectral type*; K, M, R, N, S as *late spectral type*. It is now known that variations of chemical composition from star to star are much smaller than was believed when spectra were first classified but the remaining differences are important as we shall see later, particularly in Chapter 7. Hydrogen is the most abundant element in almost all stars but it is not observed in the spectra of very hot stars because it is then ionised or of very cool stars when its spectral lines are in the ultraviolet region of the spectrum. Inside each spectral type there is a decimal subdivision, from B0 to B9 for example, with the highest surface temperature corresponding to B0. The Sun is a G2 type main sequence star.

Luminosity criteria

I can now ask if it is possible to distinguish giants and main sequence stars, for example, by studying their spectra. To do this I must ask what is the principal difference between giants and main sequence stars of the same surface temperature. It is clear from fig. 10 that the *luminosity* of a giant is very much higher than that of a main sequence star. If stars radiated like *black bodies*, there would be a simple relation between luminosity, L_s, radius, r_s and surface temperature, T_s,

$$L_s = \pi a c r_s^2 T_s^4, \tag{2.1}$$

where a is the radiation density constant and c the velocity of light. For most stars the true relation is not very different from (2.1).† If two stars have the same surface temperature but one is much more luminous than the other, it must have a

---

† Theoretical astronomers find it convenient to introduce a quantity which they call the *effective temperature*, T_e, which is defined by

$$L_s \equiv \pi a c r_s^2 T_e^4. \tag{2.1a}$$

This is the temperature of a black body with the same radius and luminosity as the star.

greater radius. Giants and supergiants are therefore very much larger than main sequence stars of the same surface temperature and that is why they are given their name; similarly white dwarfs are much smaller than main sequence stars with the same surface temperature.

The spectrum of a star is determined partly by its chemical composition but to a much greater extent by the state of *ionisation* and *excitation* of the atoms of its constituent chemical elements in the visible surface regions of the star. The state of ionisation is mainly determined by the temperature of the stellar atmosphere, but it is also affected by the density, low density favouring a higher degree of ionisation than high density (see Appendix 1). Although giants are found to have similar masses to main sequence stars, they have very much greater radii and therefore lower densities. As a result the state of ionisation of material in the atmosphere of a giant differs from that in a main sequence star with the same surface temperature. The spectral lines which are prominent in giants are slightly different from those in main sequence stars. This fact enables *luminosity criteria* to be established which make it possible to decide that a star is a giant and hence to obtain a much more accurate idea of its distance from the Sun. It is then possible to obtain an approximate idea of the distribution of stars in the Galaxy for distances which are greater than those for which parallaxes are useful. Stars are placed in luminosity classes I to V with I being the most luminous supergiants and V being main sequence stars.

Mass/luminosity relation and stellar ages

I next discuss some properties of stars which enable us to learn something about their ages. The first is the main sequence *mass/luminosity relation* (fig.14). For nearby stars on the main sequence, it is found that their luminosity is dependent on a relatively high power of their mass. The relation is not a simple power law but for a reasonable mass range a good approximation is

$$L_s \propto M_s^4. \tag{2.2}$$

Main sequence stars are believed to derive their luminosity from nuclear reactions in their central regions which convert hydrogen into helium. Because these reactions release a certain fraction of the rest mass energy of the hydrogen, the total supply of nuclear energy, E_N, satisfies

$$E_N \propto M_s. \tag{2.3}$$

As a result the total time for which a star can remain as a main sequence star, t_{ms}, satisfies the approximate relation

$$t_{ms} \propto M_s^{-3}. \tag{2.4}$$

This means that the star's lifetime as a main sequence star is a rapidly decreasing function of stellar mass. Actual values of main sequence lifetimes obtained from stellar evolution calculations are shown in Table 2.

Because the amount of energy which is released by conversion of helium to heavier elements is much less than that released by conversion of hydrogen into

Table 2. *Approximate main sequence lifetimes (in years) for stars of different masses.*

$M/M_\odot$	Lifetime
15.0	1.0×10^7
9.0	2.2×10^7
5.0	6.8×10^7
3.0	2.3×10^8
2.25	5.0×10^8
1.5	1.7×10^9
1.25	3.0×10^9
1.0	8.2×10^9
0.75	2.4×10^{10}
0.5	1.2×10^{11}

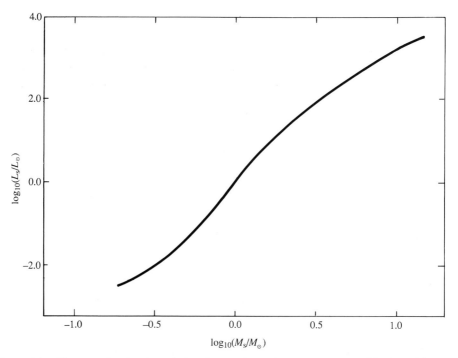

Figure 14.   The mass/luminosity relation for nearby main sequence stars. The logarithm of the luminosity is plotted against the logarithm of mass.

helium, the main sequence lifetime of a star is not very different from its total lifetime. This means that, for example, any massive stars which were formed when the Galaxy was young (10^{10} years or more ago) will have completed their life histories long ago. If I restrict my attention to main sequence stars, the figures given in Table 2 are approximate upper limits to their ages. These upper limits are very valuable for stars of early spectral types, O, B, A. For example O and B stars which are visible today must have been formed very recently in galactic history. In contrast the upper limit does not tell us anything useful about individual main sequence stars of classes G, K and M, say. A main sequence star of one of these spectral types, which was formed when the disk of the Galaxy itself was formed (about 10^{10} years or somewhat longer ago), would still be a main sequence star today. I can, however, look at this property in a slightly different way. Because such main sequence stars have not yet evolved from their main sequence state, the total population of G, K and M dwarfs† today tells us something about the total number of low mass stars which have been formed throughout galactic history. We shall find that this information about stellar ages is very important later in the book, particularly when I discuss the chemical evolution of the Galaxy in Chapter 7.

The position and motion of the Sun

Following this brief summary of stellar properties, I return to the positions and motions of the stars in the solar neighbourhood, knowing that I can distinguish between giants and dwarfs with reasonable accuracy and that I can then estimate their distances. Observations indicate that most of the stars in the solar neighbourhood and indeed in the entire Galaxy are confined to a thin disk. This is illustrated in fig. 15. The plane of the Galaxy is well-defined by the Milky Way, with the centre of the Galaxy being in the direction of the densest star clouds in the constellation Sagittarius. Because the Milky Way does define a *great circle* in the sky, it is obvious that the Sun is at present very close to the *galactic midplane* and present estimates indicate that it deviates from the midplane by no more than about 10 pc. Although most stars in the solar neighbourhood lie in a thin disk, the apparent thickness of the disk depends on the spectral type of the stars being studied. The results of the observations are shown in Table 3. It can be seen from this table that early type stars, which are necessarily young, are confined to a thinner disk than late type stars, which are of greater average age.

Although the Sun is at present near the galactic midplane, it is certainly not at rest there, as it must be moving under the gravitational attraction of all the other stars. How can I determine in what direction the Sun is moving and with what speed? The technique adopted first requires us to recognise that a star moving with typically observed stellar speeds would cross the galactic disk in no more than

---

† Main sequence stars are often called dwarfs because they are much smaller than giants.

Table 3. *Approximate half thickness of the galactic disk for stars of different types and for the interstellar matter. The last two entries refer to halo objects rather than disk objects.*

Type of object	Half thickness of disk/pc
O type stars	50
B type stars	60
A type stars	115
F type stars	190
G dwarfs	340
K dwarfs	350
M dwarfs	350
G giants	400
K giants	270
Interstellar gas and dust	100
High velocity stars	3000
Globular clusters	4000

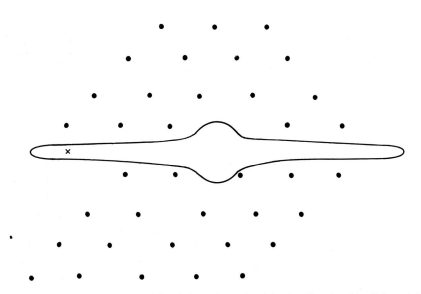

Figure 15.   A schematic view of the Galaxy from the side showing the thin disk and the central nuclear bulge. The filled circles are globular star clusters and the Sun is marked **X**.

a few times 10^7 years which is very much less than the age of the Galaxy. As a star moves away from the midplane it is attracted back by the other stars and stars must have oscillated up and down across the galactic disk many times during galactic history.† The same must be true of motion in and out in the radial direction in the disk. This suggests that the stellar system should be in a statistically steady state so that there should, for example, be as many stars moving away from the galactic midplane as towards it. This should be apparent if we look at the system of stars from an appropriate frame of reference. As the Sun cannot be expected to be at rest in this frame of reference, we may expect to observe an asymmetry of stellar velocities about the Sun and that is what is found.

The Local Standard of Rest

What I can now do is to define what is called the *Local Standard of Rest* (LSR), about which the distribution of stellar velocities is as symmetrical as possible, and I can further find the velocity of the Sun with respect to this Local Standard of Rest. If I take as complete a sample of stars as possible in the solar neighbourhood and suppose that the ith star has velocity components (u_i, v_i, w_i) in some frame of reference at the Sun in which the Sun itself is at rest, I can form the mean velocity of the local system of stars relative to the Sun

$$(\overline{u}, \overline{v}, \overline{w}) = (\Sigma u_i/N, \Sigma v_i/N, \Sigma w_i/N),‡ \qquad (2.5)$$

where N is the total number of stars included and Σ denotes the sum over the stars. The Local Standard of Rest is then assumed to have this mean motion relative to the Sun and the *solar motion* with respect to the Local Standard of Rest is

$$(u_\odot, v_\odot, w_\odot) = -(\overline{u}, \overline{v}, \overline{w}). \qquad (2.6)$$

The solar motion and random stellar velocities

If this procedure is followed through two important results are apparent. The first property is related to the distribution of stellar random velocities $(u_i - \overline{u}, v_i - \overline{v}, w_i - \overline{w})$. These are not distributed isotropically about the Local Standard of Rest. I have not at present defined a particular coordinate system. I now choose a cylindrical polar coordinate system $\bar{\omega}, \phi, z$ shown in fig. 16. In this coordinate system the z direction is perpendicular to the galactic midplane and $\bar{\omega}$ is the radial distance from the axis of symmetry through the centre of the Galaxy. If this is a coordinate system in which the LSR is at rest and $v_{\bar{\omega}}, v_\phi, v_z$ are velocities of stars relative to the LSR, it is found that

$$\langle v_\phi^2 \rangle \approx \langle v_z^2 \rangle \approx 0.4 \langle v_{\bar{\omega}}^2 \rangle, \qquad (2.7)$$

---

† These stellar oscillations will be discussed on page 107
‡ As described here the observations of all stars have been given equal weight. Because some observations are clearly more uncertain than others it is common to give higher weight to what are believed to be the best observations.

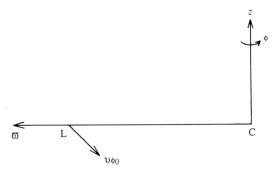

Figure 16.   The cylindrical polar coordinate system $\bar{\omega}$, ϕ, z used in describing the structure of the Galaxy. C is the centre of the Galaxy, L is the local standard of rest and $v_{\phi0}$ is the velocity of the local standard of rest about the galactic centre.

where the symbols <> indicate an average. In fact, as I shall explain shortly the Galaxy as a whole is rotating and the Local Standard of Rest has a velocity $v_{\phi0}$ in a non-rotating frame. It is more convenient to use that frame and to replace (2.7) by

$$<(v_\phi - v_{\phi0})^2> \approx <v_z^2> \approx 0.4<v_{\bar{\omega}}^2>. \tag{2.8}$$

This indication that stars have larger random speeds in the radial direction than in the other two directions will be found to be very important in Chapter 4. An estimate of the solar motion in this coordinate system is

$$v_{\bar{\omega}\odot} = -10.4 \text{ km s}^{-1}, \, v_{\phi\odot} = 14.8 \text{ km s}^{-1}, \, v_{z\odot} = 7.3 \text{ km s}^{-1}. \tag{2.9}$$

I shall use these values in discussing the motion of the Sun in the Galaxy although other investigations have given somewhat different values for the magnitude and direction of the solar motion.

High velocity stars

The second property of the determination of solar motion is that, if I consider either only stars of early spectral type or only stars of late spectral type, I obtain somewhat different values for the solar motion. Another answer is obtained if only those stars with a very high velocity with respect to the Sun are considered; these are given the name of *high velocity stars*. The early type stars are, as we have seen on page 24, young stars while the late type stars include stars of all ages. This indicates that the Sun has a different motion relative to stars of different ages which means that in some sense I must consider the Galaxy to consist of more than one system of stars. The high velocity stars are believed to be very old stars as I now explain.

When the motions of the stars in the solar neighbourhood are studied, it is found that all of the very young stars have comparatively small velocities (10–20 km s^{-1}) relative to the Local Standard of Rest and that they are therefore confined to a thin disk but that a small proportion of the late type stars, whose age cannot immediately be established, have velocities of up to several hundred km

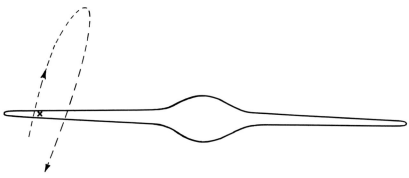

Figure 17. The orbit of a halo star showing that it passes through the disk of the Galaxy. The Sun is marked **x** and it is at rest in the frame of reference used.

s^{-1} relative to the LSR. With such velocities these high velocity stars are capable of travelling many kpc from the galactic plane before their motion is reversed by the gravitational field of the Galaxy. This means that they spend most of their time in the *halo* of the Galaxy which also contains globular star clusters as is shown in fig. 15. This illustrates one very important property of the Galaxy, which I shall discuss in much more detail in Chapter 4. A galaxy is a dynamical system in which all of the stars are moving in the gravitational attraction of all the other stars. All stars that are in the halo must pass through the disk periodically, as is shown in fig. 17 and the stars which are in the solar neighbourhood at a given time will not always remain there. Although we cannot travel to the halo to study the halo population of stars, we are able to study it through those halo stars which pass close to us. In this way we are perhaps more fortunate than if we were ourselves on a planet around a halo star; we should only be able to observe disk stars closely if we happened to be alive at the time that our Sun made one of its periodic passages through the galactic disk. In some compensation, from our viewpoint in the halo, we should have a much clearer picture of the overall structure of the Galaxy.

Variation of stellar chemical composition with age and position

It is believed that the high velocity stars form part of a halo population which contains particularly the globular star clusters. In fact, it is known that stars escape from star clusters during their life history and it is *possible* that essentially all of the individual high velocity stars were once cluster members. They may have been formed inside clusters and have subsequently escaped. A study of the Hertzsprung–Russell diagrams of globular star clusters enables their ages to be estimated and a typical age of a globular cluster is found to be 1.5×10^{10} yr. Although this value is uncertain perhaps by up to 3×10^9 yr, globular clusters are the oldest systems of which we have knowledge in the Galaxy. The high velocity stars are probably of similar age. The globular clusters and high velocity stars also have a much lower mass fraction of elements other than hydrogen and helium in their chemical composition than most disk stars. Disk stars have typically between

$\frac{1}{2}$ and 2 times the heavy element content of the Sun, which itself has about 2 per cent of its mass in elements other than H and He. In contrast, stars in some globular clusters and some high velocity stars have 1/200 the heavy element content of the Sun or even less.

I shall defer any serious discussion of these relations between stellar chemical composition, stellar age and position in the Galaxy until Chapter 7 and will simply say now that they are important properties to be explained in any study of galactic evolution. One further point about globular clusters should, however, be stated. Although I talk about them as halo objects, they do form an approximately spherical system concentric with the Galaxy and some of them never travel more than a few kpc from the galactic centre. If the chemical composition of globular clusters is considered in terms of the maximum distance that they can penetrate into the halo, it is found that the heavy element content is on average higher for those clusters that are confined close to the centre of the Galaxy. One further comment on stellar chemical composition is also appropriate. It is often supposed that the initial chemical composition of the Galaxy was devoid of heavy elements, so that the first stars formed should have been similarly devoid. Although stars are found with very low heavy element content, none have been found with no heavy elements. Could this be because they do not exist or could they be hiding far out in the galactic halo? The point that I wish to make immediately is that, because halo stars must pass regularly through the disk, we can learn something about their possible number from our failure to detect them. I shall make further comments on this point in both Chapters 5 and 7.

The rotation of the Galaxy

I have already mentioned on page 28 that the Galaxy is rotating so that the Local Standard of Rest moves about the centre of the Galaxy. I now wish to discuss the character of this rotation. The rotation of the Galaxy was discovered in the middle 1920s and it was soon realised that its existence was crucial for an understanding of the high degree of flattening of the galactic disk. It is, of course, always possible to choose a coordinate system in which the LSR does not rotate about the galactic centre and I should therefore explain what I mean by saying that the galaxy *does* rotate. I should in fact ask whether the galaxy rotates in a frame in which the observable Universe has no angular momentum. I cannot answer that question precisely but I can consider the motion of the LSR relative to a group of galaxies in the neighbourhood of our Galaxy and I shall describe the result obtained shortly. There is another trivial way in which I can say that the Galaxy must be rotating. As we shall see below, the angular velocity varies with distance from the galactic centre so that there is no rigid frame of reference in which the whole Galaxy is non-rotating. I now consider two ways in which the rotational velocity of the LSR can be estimated. They are both rather uncertain but they indicate that the LSR is moving around the centre of the Galaxy with a speed of order 250 km s^{-1}. This value of $v_{\phi 0}$ as introduced in equations (2.8) is very much greater than the speed of the Sun and most nearby stars with respect to the LSR.

The two methods, which are simple in conception if not in execution, depend on observations of globular clusters and of the Local Group of Galaxies.

The system of globular clusters is not highly flattened. As I have already mentioned and will discuss further later, we associate the flattening of the disk with its rapid rotation speed. This suggests that the system of globular clusters may be either non-rotating or slowly rotating. Let me first suppose that it is non-rotating. It is in fact observed that the system of globular clusters rotates relative to the LSR and I can now explain this observation by saying that the LSR is itself rotating about the galactic centre and moving through a non-rotating system of globular clusters. Several different observers have used observations of globular clusters to obtain estimates of $v_{\phi 0}$. They have obtained

$$\left.\begin{array}{l} v_{\phi 0} = 200\text{--}225 \text{ km s}^{-1}, \\ v_{\phi 0} = 212 \pm 16 \text{ km s}^{-1}. \end{array}\right\} \qquad (2.10)$$

The results (the first representing two independent studies) suggest that the rotational velocity is between 200 and 220 km s^{-1}.

As I shall explain in the next chapter, our Galaxy is one of a system of galaxies which is called the *Local Group of Galaxies*. It is the second most massive member of the group, the most massive being the great galaxy in Andromeda (M31). I can now suppose that the local group is non-rotating and use its apparent rotation velocity to estimate the rotation speed of the LSR. Again two different observers have obtained essentially identical estimates of $v_{\phi 0}$:

$$\left.\begin{array}{l} v_{\phi 0} = 300 \pm 25 \text{ km s}^{-1}, \\ v_{\phi 0} = 292 \pm 32 \text{ km s}^{-1}. \end{array}\right\} \qquad (2.11)$$

The results from the globular clusters and the Local Group of Galaxies do not agree but they do suggest that the velocity of the LSR may be between 200 and 300 km s^{-1} and that neither the system of globular clusters nor the Local Group of Galaxies is rotating very rapidly. What is particularly interesting is that both results are consistent with the velocity being perpendicular to the line to the galactic centre and in the galactic plane, which is what I would expect of a rotation velocity. I shall explain later that other methods suggest that the velocity is probably about 220 km s^{-1}. This implies in turn that the system of globular clusters is effectively non-rotating.

Galactic rotation in the solar neighbourhood

These observations indicate that the Galaxy is rotating, but they only give direct information about the rotation of the Galaxy in the immediate neighbourhood of the Sun. We can do little more than that if we confine our study to the stars because, as I have already stated in Chapter 1, the interstellar dust makes it very difficult for us to study the properties of the stars at large distances. Fortunately we can also study galactic rotation from observations of the interstellar gas, and I shall describe this later in the chapter. Before doing so, I show how some

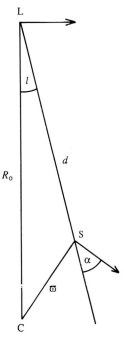

Figure 18.   The position of the Local Standard of Rest, L, and a star, S, relative to the centre of the Galaxy C. The arrows indicate the circular motions of both L and S,

information about the local variation of rotation velocity with distance from the centre of the Galaxy can be obtained from a study of the relatively nearby stars.

For most stars in the disk, their random speeds relative to the LSR are very small compared to the estimated speed of the LSR about the galactic centre. It is therefore reasonable to assume as a first approximation that the stars move in pure rotation about the centre of the Galaxy, thus neglecting both their random motions in the galactic plane and their motion perpendicular to the plane. I shall discuss the extent of these departures from pure circular motion towards the end of Chapter 4. Let us, therefore, suppose that at distance $\tilde{\omega}$ from the centre of the Galaxy the angular velocity of the stars is $\omega(\tilde{\omega})$ and that the distance of the LSR (L) from the galactic centre (C) is R_0 and its angular velocity is ω_0 (fig. 18). I call the corresponding rotation speed of the stars $v_{\mathrm{circ}}(\tilde{\omega})$ and of L, $v_{\mathrm{circ}}(R_0) \equiv v_{\varphi 0}$.

Consider the motion of a star S in the direction of galactic longitude l, which is defined in the figure. This star is intended to be quite near to the Sun at L,† but the distance is exaggerated to make the diagram clearer. Observed from L, S will have

---

† Because the LSR has been defined in terms of the motion of the stars observed from the Sun, its position is instantaneously coincident with that of the Sun. Because its velocity is different, they will not remain coincident but at any later time we can define another LSR which coincides with the Sun. Here we are only discussing the structure of the Galaxy at a single time.

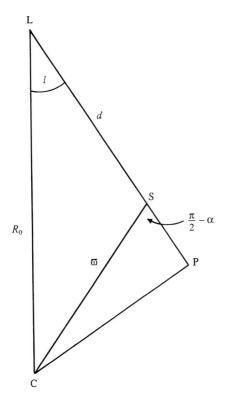

Figure 19.

a radial velocity v_R and a tangential velocity v_T, or equivalently a proper motion μ. As explained in Chapter 1 it is μ rather than v_T which can be observed. Note that by radial velocity I mean motion towards or away from the observer at L, while tangential velocity is motion across the sky. I can easily write down v_R and v_T in terms of ω, ω_0, l and d, the distance of S from L. Thus, initially

$$v_R = \omega(\tilde{\omega})\tilde{\omega} \cos\alpha - \omega_0 R_0 \sin l, \tag{2.12}$$

$$v_T = \omega(\tilde{\omega})\tilde{\omega} \sin\alpha - \omega_0 R_0 \cos l. \tag{2.13}$$

It is now easy to eliminate α. From fig. 19 it can be seen that

$$CP = R_0 \sin l = \tilde{\omega} \cos\alpha, \tag{2.14}$$

and

$$LP = R_0 \cos l = d + \tilde{\omega} \sin\alpha. \tag{2.15}$$

If (2.14) and (2.15) are used in (2.12) and (2.13), I have

$$v_R = (\omega - \omega_0)R_0 \sin l, \tag{2.16}$$

$$v_T = (\omega - \omega_0)R_0 \cos l - \omega d. \tag{2.17}$$

Oort's constants

These formulae do not depend on S being close to L. If it is I can expand ω in a Taylor series about ω_0. Thus for S near to L I write

$$\omega - \omega_0 \approx (d\omega/d\tilde{\omega})_{R_0}(\tilde{\omega} - R_0) \qquad (2.18)$$

and in addition

$$R_0 \approx \tilde{\omega} + d \cos l. \qquad (2.19)$$

If (2.18) and (2.19) are substituted into (2.16) and (2.17),

$$v_R = -\tfrac{1}{2}R_0(d\omega/d\tilde{\omega})_{R_0} d \sin 2l, \qquad (2.20)$$

$$v_T = -R_0(d\omega/d\tilde{\omega})_{R_0} d \cos^2 l - \omega d$$

$$= -\tfrac{1}{2}R_0(d\omega/d\tilde{\omega})_{R_0} d \cos 2l - [\omega + \tfrac{1}{2}R_0(d\omega/d\tilde{\omega})_{R_0}]d. \qquad (2.21)$$

Equations (2.20) and (2.21) are usually written

$$v_R = Ad \sin 2l, \qquad (2.22)$$

$$v_T = Ad \cos 2l + Bd, \qquad (2.23)$$

where A and B are called *Oort's constants* of galactic rotation. They are normally written in terms of the local rotation velocity and its derivative. In terms of these

$$A = \frac{1}{2}\left[\frac{v_{\phi 0}}{R_0} - \left(\frac{dv_{\text{circ}}}{d\tilde{\omega}}\right)_{R_0}\right], \qquad (2.24)$$

$$B = -\frac{1}{2}\left[\frac{v_{\phi 0}}{R_0} + \left(\frac{dv_{\text{circ}}}{d\tilde{\omega}}\right)_{R_0}\right]. \qquad (2.25)$$

Determination of Oort's constants

A and B are both combinations of the velocity, $v_{\phi 0}$, of the LSR, the first derivative of the circular velocity and the distance to the galactic centre, R_0. If values can be obtained for A and B, I shall need only one more observation to determine all three quantities. How can I obtain values for A and B? Let us first consider stars of known distance whose radial velocities are observed. From (2.22) I have

$$v_R/d = A \sin 2l, \qquad (2.26)$$

which implies that, if I plot v_R/d against l, a sine curve should result whose amplitude is A. There are, however, three complications in the real Galaxy in which stars do not move in purely circular orbits. The first is that we observe radial velocities with respect to the Sun and not with respect to the LSR. If we are

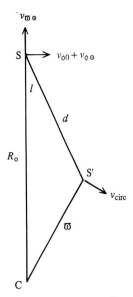

Figure 20. The solar motion.

observing stars in the galactic plane and the components of solar motion in the plane are $v_{\bar{\omega}\odot}$ and $v_{\phi\odot}$, I can easily modify (2.17) and (2.18) to take account of this (see fig. 20). Thus with respect to the Sun I have

$$v_{R\odot} = Ad \sin 2l + v_{\bar{\omega}\odot} \cos l - v_{\phi\odot} \sin l, \tag{2.27}$$

$$v_{T\odot} = Ad \cos 2l + Bd - v_{\bar{\omega}\odot} \sin l - v_{\phi\odot} \cos l. \tag{2.28}$$

As values of $v_{\bar{\omega}\odot}$ and $v_{\phi\odot}$ are known, the radial velocities of the stars can be adjusted so as to be with respect to the LSR and these values can be used in (2.26). Note that in a complete treatment, using stars above and below the plane, it is also necessary to introduce the solar motion perpendicular to the galactic plane. If galactic latitude b is introduced, so that it is zero in the galactic plane just as is terrestrial latitude on the equator, stars will have two components of tangential velocity and hence of proper motion corresponding to motions in the galactic plane and perpendicular to the plane. A further subtlety is that the velocity of galactic rotation may itself vary with distance from the galactic plane. I will ignore these complications in what follows.

There are two further reasons why the observations cannot be expected to give the smooth curve predicted by equation (2.26). The first is that the stars which we are observing also have deviations about a pure circular motion and the second is that even for stars of supposedly known distance there are considerable uncertainties in the distance. To a first approximation these two effects should lead to a random scatter rather than to a systematic effect of the type produced by the solar motion and a fairly smooth sine curve can be drawn through the mean of the

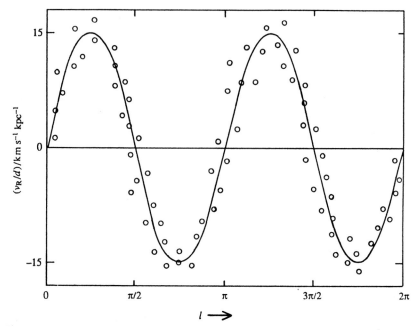

Figure 21.   The determination of A from radial velocities. A schematic plot of v_R/d against l together with the sine curve from which A can be determined. (Note, these are not actual observations.)

observations (fig. 21). This method has been used by many investigators with an average of all their results giving

$$A \approx 14.5 \text{ km s}^{-1} \text{ kpc}^{-1}.\dagger \tag{2.29}$$

If I next consider stars of known proper motion, which is essentially v_T/d, I see that in the simple model

$$v_T/d = A \cos 2l + B. \tag{2.30}$$

It is frequently stated in books on galactic structure that equation (2.30) implies that observations of proper motions even without knowledge of stellar distances should be capable of giving values of both A and B. It is obvious that this statement is not strictly correct when it is realised that proper motions are observed with respect to the Sun. If equation (2.28) is divided by d, it is clear that the correction to the proper motion depends on d. A large number of studies of the values of A and B from proper motions have been carried out. Proper motions are less accurate than radial velocities and the divergence between different results is quite great. However the average value of A is almost exactly the same as that given by radial velocities, while the mean value of B is

$$B \approx -12 \text{ km s}^{-1} \text{ kpc}^{-1}. \tag{2.31}$$

$\dagger$ The units of A are rather mixed, A being a frequency, but they are convenient in use. A natural unit of velocity is km s^{-1} and a kpc is a typical distance inside a galaxy.

The uncertainty in both A and B from proper motions is about $3 \text{ km s}^{-1} \text{ kpc}^{-1}$. In due course it is hoped that more and improved proper motions, such as those being obtained by the HIPPARCOS satellite, will lead to better estimates of both A and B. Although the average values of A and B give a rotation velocity which is decreasing outwards at the solar radius, a flat or increasing velocity cannot be ruled out. I shall explain that this is important when I consider galactic mass models in Chapter 5.

I have already mentioned in Chapter 1 one method of determination of R_0. This is Shapley's original method which assumes that the centre of the Galaxy is at the centre of the system of globular clusters. There are three problems in this technique. There must be a sufficiently complete catalogue of clusters evenly distributed about the galactic centre. There must be a well-defined absolute luminosity for a type of cluster star, so that distance can be deduced from apparent luminosity. The distance estimates must not be too seriously affected by absorption of starlight. Two recent investigations have given

$$\left. \begin{array}{l} R_0 = 8.5 \pm 1.0 \text{ kpc}, \\ R_0 = 6.8 \pm 0.8 \text{ kpc}. \end{array} \right\} \tag{2.32}$$

A second method involves the observation by Baade that, although extinction of starlight towards the galactic centre is very great, there are lines of sight which pass quite close to the centre where extinction is anomalously low (Baade's windows). If a class of stars known as RR Lyrae variables is observed, their region of maximum density is assumed to be close to the galactic centre and, if their absolute magnitude is assumed known from other studies, an observation of their apparent magnitude provides a distance to the galactic centre. The principal problems here are related to disagreements about the absolute luminosity of RR Lyraes and corrections for the extinction which actually occurs. Three results obtained are

$$\left. \begin{array}{l} R_0 \approx 8.7 \text{ kpc}, \\ R_0 = 7.95 \pm 0.69 \text{ kpc}, \\ R_0 = 6.94 \pm 0.58 \text{ kpc}, \end{array} \right\} \tag{2.33}$$

where the second and third figures come from the same data but with different assumptions about RR Lyrae luminosities. A final direct method studies the group of stars known as Mira variables in Baade's window and gives

$$R_0 = 8.3 \pm 0.6 \text{ kpc}. \tag{2.34}$$

Another way of estimating R_0 is as follows. Stars which are also at distance R_0 from the galactic centre have no radial velocity with respect to the LSR if random stellar velocities are neglected. For such stars, it is easy to see from fig. 18 that

$$d = 2R_0 \cos l, \tag{2.35}$$

so that if d and l are known R_0 can be found. Observations of stars with zero radial velocity have given $R_0 \approx 9 \text{ kpc}$ and $R_0 \approx 10.4 \pm 0.7 \text{ kpc}$. If I use the values of A

and *B* obtained from nearby stars and an average of all the values of R_0, it is easy to see that they are consistent with the earlier estimate

$$v_{\phi 0} \approx 220 \text{ km s}^{-1}. \tag{2.36}$$

Further information about the rotation of the Galaxy is obtained from study of the interstellar gas and I defer this until I have given a general discussion of the interstellar medium.

The interstellar medium

The existence of an interstellar medium is suggested by a variety of observations, some of which have been known for a very long time. These include the existence of dark patches in dense star clouds (fig. 22) which are presumably caused by obscuring matter between us and the stars. The first really convincing demonstration of the existence of a general interstellar medium came in the 1920s. It was found that in certain directions in the sky many stars had absorption lines which appeared at the same wavelengths in contrast to the main spectra of the stars which were shifted to the red or blue by varying amounts depending on the velocities of the stars. This observation could only be understood if there was absorbing matter which produced the lines between the stars and the observer (fig. 23). Those spectral lines could be interpreted as being due to elements such as calcium and sodium in their neutral (un-ionised) form. In the years that followed it became apparent that the chemical composition of most stars was dominated by the element hydrogen, as I have already mentioned. It seemed likely that the interstellar medium would share the same chemical composition, especially as it is believed that stars have formed and are forming out of interstellar gas. Indeed there are some clouds of gas which surround hot stars and which are raised to a high temperature and spectral lines of hydrogen are emitted by these hot clouds. It was, however, for a very long time difficult to see how any hydrogen which was associated with the absorbing clouds could be detected even if it were the most important constituent of the clouds. The reason for this is quite simple. The clouds must be cold otherwise the calcium and sodium would not be neutral. Cold hydrogen possesses no absorption lines in the visible region of the spectrum. If hydrogen is in its ground state even excitation from the ground state to the first excited state requires ultraviolet radiation (fig. 24) and this ultraviolet region of the spectrum cannot be studied with ground-based telescopes. The development of satellite-based ultraviolet telescopes, such as the International Ultraviolet Explorer (IUE), has changed this situation, but first studies of interstellar hydrogen came through radio astronomy. Ultraviolet studies have been particularly important in detecting molecular hydrogen which does not possess a radio spectral line corresponding to that of atomic hydrogen, which I now discuss.

Figure 22.   The dark mass of absorbing dust and gas familiar to southern hemisphere observers as *the Coalsack*. The bright image at the left is the greatly over-exposed image of α Crucis, the brightest star in the Southern Cross. (Photograph taken on the UK 1.2 m Schmidt Telescope, reproduced by permission of the Royal Observatory, Edinburgh.)

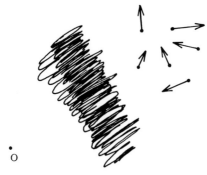

Figure 23.   Interstellar absorption lines. If the observer at O studies the stars whose random motion is shown, through the dense gas cloud, absorption lines due to the cloud will occur at the same wavelengths in the spectra of all of the stars, whereas the stellar lines will be shifted to a variable extent by the Doppler effect.

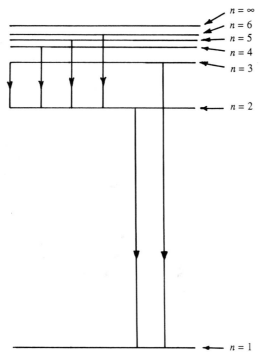

Figure 24.   The electron energy levels of the hydrogen atom with the lower transitions of the Lyman series (to $n = 1$) and the Balmer series (to $n = 2$) marked.

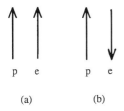

(a) (b)

Figure 25.   The possible arrangement of electron and proton spin in the hydrogen atom.

The 21 cm line

The problem of the detection of the cold interstellar hydrogen was solved by the advent of radio astronomy and by the realisation that even cold hydrogen can absorb and emit waves at a specific wavelength of 21 cm† (0.21 m). This arises because the proton and electron, which form the hydrogen atom, both possess a spin and a magnetic moment. Because they behave like small dipoles, there is an energy difference according as the spins are parallel or antiparallel, with the parallel state being more energetic (fig. 25). If the spins change from being parallel to antiparallel, radiation of a wavelength of 21 cm is emitted. Note that this energy difference between the two components of the split ground state is only 6×10^{-7} of the difference between the ground state and the first excited state, which accounts for the very much longer wavelength of the radiation.

The existence of this 21 cm radiation and its possible importance in astronomy were predicted by the Dutch astronomer H. C. van de Hulst in the middle of the Second World War. As soon as serious astronomy could be resumed, several groups in the world built radio receivers capable of receiving the radiation, which was detected in 1951. It immediately gave very valuable information about the structure of the Galaxy and in particular about galactic rotation. We shall see that the interpretation of the observations is not completely unambiguous because we have no direct way of determining the distance to an object which emits radio waves. Thus, the interpretation is based on assumptions which may be wrong in detail. I start with the assumption which I have already made for the stars, which is that the gas moves with purely circular velocity, apart from thermal motion of its atoms.

Galactic rotation from radio observations

When we look in the direction of a given galactic longitude l, the radio spectrum in the region of 21 cm may look as shown in fig. 26. The radiation does not all appear at 21 cm because of Doppler shifts which arise both because the gas as a whole is moving towards us or away from us and because of the random velocities of the hydrogen atoms inside any particular cloud of gas. Thus in fig. 26 we can see evidence for the existence of several distinct clouds of gas in the

---

† Because this is always called 21 cm radiation, I do not use SI units for it.

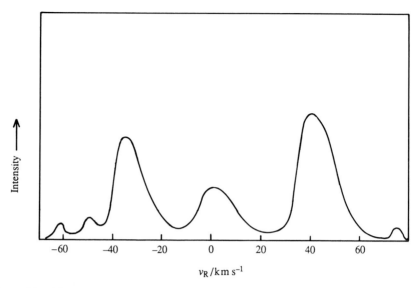

Figure 26.   The appearance of radiation close to 21cm in the direction of galactic longitude l. The observed wavelengths have been converted to velocities of approach or recession. The peaks are interstellar clouds while the breadth of the peaks is caused by random motions inside clouds.

direction of galactic longitude l. Each peak in the spectrum can probably be associated with one cloud and the width of the peak is produced by random velocities in the cloud. The clouds will be at different distances from the LSR.

Now consider what determines the radial velocity of a cloud if it is moving in purely circular motion. Formula (2.16)

$$v_R = (\omega - \omega_0)R_0 \sin l \qquad (2.16)$$

applies. Suppose now that $\omega(\tilde{\omega})$ decreases outwards from the axis of rotation. We shall see later that this is true both for our Galaxy and for other galaxies; the values of A and B which have been found indicate that it is true in the solar neighbourhood. Then from (2.16) it is possible to see that, as we move in the direction of galactic longitude l towards the inner regions of the Galaxy, $\omega - \omega_0$ and hence v_R increases until the point P shown in fig. 27 is reached. After that the radial velocity decreases again until it vanishes at L′ and subsequently becomes negative. If there were gas at all points along the line of sight, it would be possible to say that the maximum velocity observed in direction l, or more precisely the velocity of the most rapidly moving cloud, would be the velocity appropriate to P. The radial velocity of the cloud is easily obtained from the wavelength of observation by using the standard formula relating Doppler shift to velocity. If R_0 is assumed known from methods which I have described, CP is known and it is easy to deduce the angular velocity at distance CP from the galactic centre. By observations at a variety of galactic longitudes, it is then possible to build up a rotation curve for all distances less than R_0. Even if R_0 is unknown, CP is a known fraction of R_0

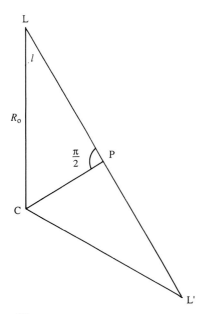

Figure 27.

($CP/R_0 = \sin l$) so that equally a rotation curve can be constructed. As there may not, however, be any gas at the potential point of maximum velocity for a given l, the value obtained will be less than or equal to the true value.

The rotation curve for the Galaxy

When this procedure is followed and is combined with the values of $v_{\phi 0}$ and its gradient obtained from study of the nearby stars, a rotation curve is constructed which has the general character shown in fig. 28. The detailed structure shown in the inner regions is rather uncertain but what appears certain is that, after a rapid rise near the centre of the Galaxy, the rotation speed is relatively slowly varying for quite a large distance. This means that the Galaxy is a differentially rotating system with the angular velocity decreasing outwards; the dip in the rotation curve after the initial rise is not enough to prevent the monotonic decline of angular velocity. I shall show that the phenomenon of differential rotation has a straightforward explanation when I consider the relation between rotation velocity and galactic mass distribution in Chapter 5, but I shall also explain that this differential rotation has some very important consequences for the spiral structure of our Galaxy and other spiral galaxies.

Although the rotation curve of the form shown in fig. 28 appears to be generally correct, the results obtained for positive and negative galactic longitude, which should be the same if the only important motion of the gas is one of rotation, do not agree in detail (fig. 29). The data shown in fig. 29 are now very old but they have not been superseded by more recent observations. It does not appear likely that the differences are entirely due to the failure of the gas to be present at the

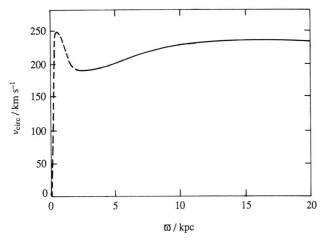

Figure 28.   A smooth rotation curve for the Galaxy. The dashed region is more uncertain than the solid region.

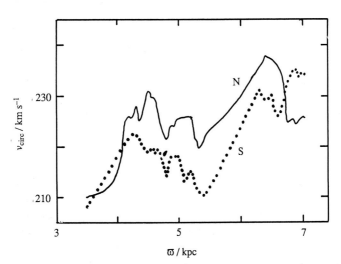

Figure 29.   Differing rotation curves found in the northern hemisphere (N) and the southern hemisphere (S).

maximum density points or to random gas cloud motions. The simplest explanation is a deviation from strict circular motion particularly in the central regions of the Galaxy. It is easy to see, for example, that, if a velocity of general expansion is added to the circular motion, this will produce an asymmetry in the radial velocity curves for positive and negative l and in the rotation curves deduced on the assumption that motions are strictly circular. It appears that gas *is* moving outwards from the centre of the Galaxy, possibly as a result of an explosion in the central regions some millions of years ago. I shall have more to say about much more violent gas motions in the central regions of other galaxies in Chapter 3.

I complete this discussion by describing how observations of the gas can be used to give a value of AR_0 and hence of R_0 if A is assumed known. I consider values of $\tilde{\omega}$ close to R_0. From equations (2.16), (2.18) and the definition of A, it is possible to see that

$$v_R = 2A(R_0 - \tilde{\omega}) \sin l. \tag{2.37}$$

From fig. 27 it is seen that v_R has its maximum value in direction l when

$$\tilde{\omega} = R_0 \sin l. \tag{2.38}$$

This value of $\tilde{\omega}$ will be close to R_0 if $l \approx 90°$. Combining (2.37) and (2.38) I have

$$v_{\max} = 2AR_0(1 - \sin l) \sin l. \tag{2.39}$$

If observations are made for l close to but not exactly equal to 90°, a value of AR_0 is obtained. Equation (2.39) can be generalised to enable observations which are not very close to $l = 90°$ to be used. Results obtained during the past thirty years have varied between 156 km s^{-1} and 103 km s^{-1}, which indicates that it is not intrinsically a very accurate method. The average of the results gives

$$AR_0 = 124.5 \pm 17 \text{ km s}^{-1}, \tag{2.40}$$

so that $R_0 = 8.6 \pm 1.2$ kpc if $A = 14.5$ km s^{-1} kpc^{-1}.

In 1964 the International Astronomical Union adopted official values of A, B, R_0 and $v_{\phi 0}$ which were $A = 15$ km s^{-1} kpc^{-1}, $B = -10$ km s^{-1} kpc^{-1}, $R_0 = 10$ kpc and $v_{\phi 0} = 250$ km s^{-1}. The intention was that, if all investigators of galactic structure used these standard values, this would remove one source of uncertainty in comparing results obtained by different authors. It subsequently became apparent that the standard Galaxy differed from the real Galaxy. In 1982 a committee was set up to recommend new standard values. Its recommendations, which were accepted in 1985, were $R_0 = 8.5$ kpc, $v_{\phi 0} = 220$ km s^{-1}. New values of A and B were not recommended, but the values of r_0 and $v_{\phi 0}$ imply that $A - B = 25.9$ km s^{-1} kpc^{-1}. This is not very different from the value of $A - B$ implied by the best values of A and B given earlier or from the older standard value of 25 km s^{-1} kpc^{-1}. The Galaxy now appears to be significantly smaller than was once believed; as was mentioned in Chapter 1, Shapley's first estimate of R_0 was 15 kpc. At the same time other galaxies are now known to be further away from us than was believed sixty years ago and as a result they are larger than was then thought. As a result, the Galaxy, which once appeared to be a supergiant galaxy, is now a rather ordinary spiral galaxy.

Distribution of the gas

The hydrogen gas is mainly concentrated in a thin disk which is comparable in thickness with the disk occupied by the young stars. The gas which is detected by 21 cm radiation is cold and this does not fill the disk uniformly but forms clouds as has already been indicated in fig. 26. There is also evidence for the existence of hotter gas of lower density between the clouds and of even hotter gas of even lower density which fills a significant fraction of the disk and is believed to have been heated by the explosions of supernovae. In addition any gas which happens to be close to stars of very high surface temperature is ionised by the ultraviolet radiation emitted by the stars and forms HII regions of the type which have already been mentioned on page 15. The clouds of cold gas are also not uniformly spaced through the disk and the positions of maximum gas density appear to delineate spiral arms which are at least in good qualitative agreement with the spiral arms outlined by the most luminous blue stars and the HII regions. Although it is quite clear that the Galaxy is a spiral galaxy, it is not at all easy to determine the fine details of the spiral structure from our own position inside the disk. For that reason I do not show a picture of the spiral structure of the Galaxy. A much better view is obtained of a galaxy such as M51 (cover picture). If this is photographed in the light of Hα, arms of the form shown in fig. 30 appear very clearly. It is very unlikely that the Galaxy has such a regular spiral structure.

In addition to the gas in the plane of the Galaxy, there is an indication that there is some gas outside the plane and also that the gas disk is rather badly buckled in its outer regions (fig. 31). I shall discuss this buckling further on page 73 in Chapter 3. Because we can determine only the velocity of the interstellar gas from its radio emission and not its distance, there are difficulties in deciding whether the high

Figure 30.   A sketch of the spiral structure of the galaxy M51, a photograph of which is shown on the cover and on page 57. The satellite irregular galaxy is apparently near the end of the spiral arm at the bottom of the diagram.

Figure 31. The gas disk of the Galaxy, showing a buckling that occurs at large distances from the galactic centre. The Sun's position is marked **x**.

latitude *high velocity clouds* that have been observed are in our Galaxy or whether they are intergalactic clouds. Even if they *are* in or near to the Galaxy, there are disagreements about whether they are clouds which have been pulled out of the plane of the Galaxy by the same phenomenon which has led to the distortion of the gas disk itself or whether they are clouds of intergalactic gas which are being accreted by the Galaxy. I shall return to this point when I discuss the Local Group of Galaxies in the next chapter. The possible continuing accretion of intergalactic gas will also enter my discussion of the chemical evolution of the Galaxy in Chapter 7.

Interstellar dust

I have mentioned several times that an important constituent of the interstellar matter is *interstellar dust* which, rather than the gas, is responsible for most of the absorption of starlight. I shall now describe briefly how the dust has been discovered and what is known about it. Light from a distant star can be both *scattered* and *absorbed* by interstellar matter. Scattering simply involves a change in direction of the radiation. In the case of absorption a photon disappears. The absorbed radiation is subsequently re-emitted but in this case as well as a change in direction there is probably a change of frequency. If absorption and scattering occurred equally for radiation of all wavelengths, a star would merely look fainter and hence seem further away than it is. Actually there is a strong wavelength dependence of the extinction with blue light being more affected than red light. Because of this, the colour of a star changes but its absorption spectral lines, formed in the star's atmosphere, do not. The star then appears to have the wrong spectral type for its colour. This fact enables an estimate to be made both of the total amount of absorption and of the variation of absorption with wavelength.

It is then necessary to postulate some absorbing agent in the interstellar medium which could produce both the correct quality and quantity of absorption. It proves quite impossible to obtain the required result with any plausible atoms or molecules, but the absorption is much easier to understand if the interstellar medium contains small solid particles whose size is approximately equal to the wavelength of light. There is as yet no complete agreement about what is the chemical composition of such *interstellar grains*. Graphite, silicates and dirty ice (ice with impurities) have all been proposed and are all capable of explaining some of the characteristics of interstellar absorption. It seems likely that several types of interstellar grain are present.

Figure 32.   Polarisation of starlight. Starlight will be polarised when it reaches the observer, O, if it has been scattered by aligned particles as shown.

Polarisation of starlight

Several other observations support the existence of interstellar grains. There is for example the observation of the *polarisation of starlight*. Light from stars is sometimes found to be significantly polarised and in most cases it is found that stars of all types in the same direction in the sky have polarised light. This indicates that it is very unlikely that the light was polarised when it was emitted, and that the polarisation has been produced by scattering in the interstellar medium. Thus, light which is initially unpolarised will become polarised if it is absorbed or scattered in such a way that light in one sense of polarisation is affected preferentially. This will happen if asymmetrical scattering particles are not orientated at random but are aligned (fig. 32). It is believed that the scattering is produced by interstellar grains which are elongated and which are partially aligned by an *interstellar magnetic field*.

Interstellar magnetic field and cosmic rays

A detailed discussion of the properties of the interstellar magnetic field and cosmic rays is deferred until Chapter 6 which is concerned with the dynamics of the interstellar medium, but some of the properties will be summarised briefly now. The *cosmic rays* which are observed at the Earth are extremely high energy charged particles which are reaching the Earth approximately equally from all directions in space. As they move almost with the speed of light, they could escape from the Galaxy into intergalactic space in a time of order 10^5 years if nothing restrained them. In Chapter 6 I discuss evidence that cosmic rays spend nearer 10^7 years in the galactic disk and that they are trapped in the disk by the action of the magnetic field. This is a second piece of evidence for the existence of a large scale interstellar magnetic field. Charged particles moving in a magnetic field radiate energy, and radio waves, which reach us from both the disk and halo of the Galaxy and which are not associated with interstellar gas clouds or other discrete sources of radio emission, are most readily interpreted as being produced by cosmic ray electrons moving in a magnetic field. This suggests that cosmic rays as well as a magnetic field are a component of the whole Galaxy.

Interstellar molecules

The most significant discovery in the interstellar medium in recent years has concerned *interstellar molecules*. These have mainly been found by the techniques of radio astronomy. Molecules tend to have large numbers of spectral lines in the *microwave* or *infrared* regions of the electromagnetic spectrum and most molecules have been discovered by means of observations at wavelengths somewhat shorter than the 21 cm line of atomic hydrogen. The first molecule to be discovered by these methods was the *radical* OH which was found to be distributed very widely in the Galaxy. Other molecules which have been identified in this manner and which have been discovered in many places in the Galaxy are carbon monoxide and formaldehyde. Water is also very widespread. The most abundant molecule is certainly H_2 but this was also only discovered much later by observations in the ultraviolet region of the spectrum. This discovery was only possible by use of ultraviolet telescopes mounted on satellites. It appears that the total amount of H_2 is probably comparable with the total amount of atomic hydrogen.

The existence of relatively small molecules was not very surprising, but the real surprise was the discovery of dense clouds of gas containing very large numbers of rather complex molecules. An (incomplete) list of the molecules discovered to date is given in Table 4 from which it can be seen that such large organic molecules as ethyl alcohol and cyanoacetylene have been found. It seems almost certain that even larger molecules will be found in due course. In contrast with the simple molecules, the complex molecules have only been discovered in a small number of dense clouds. The largest concentration of molecules discovered is very close to the centre of the Galaxy; because radio waves are only very slightly absorbed by interstellar matter, it is possible to observe molecules near the galactic centre even though the light from any star placed there would be very seriously dimmed. The formation and preservation from destruction of these complicated molecules is a very interesting problem. They will form only very slowly by successive collisions between atoms and molecules in even the most dense interstellar clouds because these are vacua by usual laboratory standards. Densities of $\sim 10^{12}$ m^{-3} and temperatures of ~ 10 K give times between collisions no shorter than 10^6 s even for two hydrogen atoms; the times are very much larger for other atoms and molecules with much lower abundances. If complex molecules are to be built up by successive collisions they must be shielded from the ultraviolet radiation from hot stars which would easily dissociate them. It seems that they are most likely to be formed and preserved in the central regions of dense clouds, which also contain dust which absorbs the starlight incident on the cloud. Such a close association between dust and molecules is observed. Because the dense clouds are much denser than the general interstellar medium, they are almost certainly not in equilibrium but are contracting under gravity. Such contraction is likely to take 10^6–10^7 yr, so that molecules have that long in which to form. There is in fact some evidence that the clouds are not contracting as quickly as might be expected. Their contraction may be resisted by a magnetic field and by bulk turbulent motions as

Table 4. *An (incomplete) list of molecules which have been discovered in the interstellar medium.*

H_2	CH_2CO	CH_3NH_2	NH_2CHO
CH	CH_3OCHO	CH_3CH_2CN	SO
C_2	$CHOOCH_3$	CH_2CHCN	SO_2
C_2H	CH_3CH_2OH	HC_3N	SN
C_3H	$HCOOH$	HC_5N	H_2S
C_4H	C_3O	HC_7N	CS
C_5H	$(CH_3)_2CO$	HC_9N	OCS
C_6H	HC_3HO	$HC_{11}N$	H_2CS
CH_3C_2H	C_5O	CH_3C_3N	$HNCS$
CH_3C_4H	NH_3	C_3N	C_2S
C_2H_2	CN	CH_3NC	C_3S
OH	HCN	CH_3CN	SiO
CO	HNC	NO	SiS
H_2O	CH_2CN	HNO	HCl
CH_3OH	NH_2CH	$HNCO$	PN
HCO	CH_2NH		

well as by the thermal pressure of the gas. The contracting dense clouds are almost certainly going to form new stars in the near future.

A large proportion of the remainder of this book is concerned with the Galaxy and in particular with what deductions can be drawn from the observations which I have discussed in this chapter. Before I continue this discussion, Chapter 3 is concerned with a brief account of the properties of different types of galaxies and of the way in which they are distributed in the Universe.

Summary of Chapter 2

In this chapter I have explained that most of the stars and gas and dust in the Galaxy are distributed in a very thin disk. There is in addition an approximately spherical distribution of globular star clusters and individual high velocity stars which is known as the galactic halo. The flattening of the galactic disk is believed to be due to rapid rotation. I discuss how observations of the apparent motions of the system of globular clusters and of the local group of galaxies, of which our Galaxy is the second largest member, can be used to estimate the rotation speed of the Sun about the centre of the Galaxy. Observations of the motions of stars in the neighbourhood of the Sun and of gas clouds at greater distances can also be used to obtain information about the rotation of the Galaxy and of the distance of the Sun from the galactic centre. In particular it is found that the Galaxy does not rotate like a rigid body; the angular velocity decreases with distance from the centre of the Galaxy. The random velocities of the stars in the solar neighbourhood can also be studied. It is found that, apart from a small number of high velocity stars, the random velocities are much smaller than the circular velocity. In addition the amplitude of stellar velocities towards and away from the galactic centre is greater than the random velocities in other directions. Observations of the chemical compositions of stars show that almost all stars are mainly composed of hydrogen and helium but that the halo stars are very deficient in heavy elements compared with the disk stars.

The interstellar gas and dust and cosmic rays also form a thin disk and the gas possesses a spiral structure which is shared by the most luminous young stars. Much of the gas is in cold interstellar clouds principally composed of atomic hydrogen, which are separated by hotter, less dense gas. Some of the gas is in dense clouds which contain many molecules, including some such as ethyl alcohol, which contain many atoms, although the principal constituent is molecular hydrogen. There appears to be a large scale magnetic field in the disk of the Galaxy. The magnetic field keeps the cosmic ray particles in the Galaxy for much longer than would otherwise be the case. If there were no magnetic field the cosmic rays would leave the Galaxy in less than 10^5 years but they are constrained to move in spiral orbits around the magnetic field lines.

3

Properties of external galaxies

Introduction: the Hubble classification of galaxies

In the preceding chapter I have discussed in considerable detail the properties of one individual galaxy – the Galaxy. In this chapter I discuss other galaxies (*external galaxies*) and I compare and contrast their properties with those of the Galaxy. The easiest property of a galaxy to discuss is its visual appearance. Soon after the existence of external galaxies had been established in the early 1920s it was realised that galaxies of regular shape could be divided into two main classes, *spiral galaxies* and *elliptical galaxies*. Subsequently it was realised that the spirals should be subdivided into ordinary spirals and *barred spirals* and that a further class known as *lenticular galaxies* should be introduced. In addition there were *irregulars*, galaxies possessing no obvious symmetry. In the 1930s Hubble introduced his classification of galactic types which, with some modifications, is still used today.

The simplest version of the Hubble classification is illustrated in fig. 33. At the time that Hubble introduced the classification, he thought that it might represent an evolutionary sequence with galaxies possibly evolving from elliptical to spiral form but, as we shall see later, that is not believed to be true today. There are alternative classifications of galaxies in use but the Hubble system is essentially adequate for the present book. There is one important group of galaxies which was not known to Hubble because its members are rare and the nearest one is a very large distance from the Galaxy. These galaxies are known as cD galaxies. They are giant ellipticals and are usually the most luminous galaxies in rich clusters of galaxies. Frequently they have complex or multiple nuclei and it is believed that, as will be mentioned again later, this may arise from galaxy mergers. I now give further details of the visual appearance of different types of galaxy without at present relating visual appearance to physical properties.

52

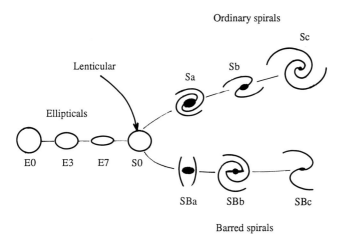

Figure 33.   The Hubble classification of galaxies. The ellipticals are shown edge-on and the lenticulars, spirals and barred spirals are shown face-on.

Elliptical galaxies

The elliptical galaxies are classified according to the ratio of their apparent major and minor axes. Thus, if a and b are these apparent (semi) axes, a galaxy is labelled E_n, where

$$n = 10(a - b)/a. \tag{3.1}$$

For no elliptical galaxy is n observed to be greater than 7. Although the observed value of n will always be less than or equal to the true value because of projection effects and a spheroidal galaxy viewed down its minor axis will always appear circular, the sharp cut-off at the observed value of n equal to 7 indicates that there really cannot exist spheroidal or ellipsoidal galaxies with a high degree of flattening. This is in contrast with all other regularly shaped galaxies (lenticular, spiral and barred spiral) which are very highly flattened like the Galaxy. Until a little over ten years ago it was believed that all elliptical galaxies were oblate spheroids with the cause of their relatively low degree of flattening being a rotational velocity which was comparable with the random velocities of the stars. It is now clear that many elliptical galaxies are triaxial. The evidence for this statement is as follows.

The extreme flattening of spiral galaxies is attributed to rapid rotation. When all elliptical galaxies were believed to be spheroidal, it was assumed that their modest flattening was also caused by much slower rotation. It is not easy to measure the rotation of elliptical galaxies, as will be mentioned later, but when it was measured for some galaxies it was found that the rotation was too slow even to cause the observed degree of flattening. As a result it was not clear that the rotation would define an axis of symmetry for the galaxy. A more direct demonstration of triaxiality comes from the observation of what is known as

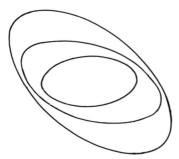

Figure 34.   In some elliptical galaxies the contours of constant light intensity rotate relative to one another as shown. (The effect is exaggerated.) This cannot happen in a spheroidal galaxy.

rotation of isophotes. An isophote is a contour of constant light intensity. If a galaxy is spheroidal, all the isophotic surfaces will be spheroids and, if the spheroid is observed from any direction, the projected isophotes will be ellipses with their major axes parallel. This is not necessarily so if the galaxy is triaxial with the relative sizes of the three axes varying between the central regions of the galaxy and the surface. In that case a rotation of isophotes as shown schematically in fig. 34 can occur and it was an observation of this which first clearly demonstrated that some galaxies were triaxial. Photographs of elliptical galaxies with two different values of n are shown in figs. 35 and 36.

Spiral, barred spiral and lenticular galaxies

The spiral galaxies were the first type of galaxy to be discovered, because the most luminous galaxies close to the Galaxy are spirals. They are so named because of the approximately spiral distribution of their light which can be seen in the photograph of a spiral galaxy in fig. 37. Originally it appeared that the spiral galaxies were far from being axisymmetric but it is now realised that the mass distribution is much less asymmetric than the light distribution. The light is mainly produced by massive young stars which are concentrated in the regions of the spiral arms. These stars are either seen directly or their presence is deduced from the observation of hot gas clouds (HII regions) which radiate strongly in visible light and which are themselves kept hot by the stars embedded in them, whose radiation is principally in the ultraviolet. In contrast, the mass is mainly in the form of older low mass stars which are essentially uniformly distributed across the disk. Spiral galaxies are highly flattened, at least in their light distribution, and that is illustrated in fig. 38.

The main difference between barred spirals and spirals is that the central nucleus, which is approximately spherical in the case of ordinary spirals, is distinctly elongated to form a *bar*. Because the nucleus always contains a significant amount of mass, the barred spirals are certainly asymmetrical. As shown in fig. 33, ordinary spirals have sub-classifications Sa, Sb, Sc and barred spirals SBa, SBb, SBc. Along both sequences, the classification a, b, c, refers both

Figure 35.   An E0 galaxy. The giant elliptical galaxy M87 in Virgo. (Photograph from the Hale Observatories, Palomar 5 m telescope.)

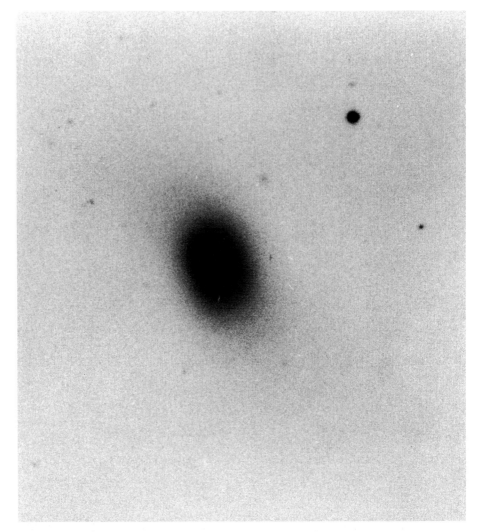

Figure 36.   An E4 galaxy. NGC 1389 in the Fornax cluster of galaxies, a negative print. (Photograph taken on the 3.9 m Anglo-Australian Telescope reproduced by permission of the Anglo-Australian Telescope Board.)

to the size of the nucleus and to the tightness of the spiral structure. The nucleus becomes smaller and the arms less tight as one goes from a to c. Since Hubble's original classification galaxies known as Sd and SBd have been added. As more detailed observations are made of ordinary spiral galaxies, it appears that many of them may have small triaxial bars in their central regions which makes the distinction between S and SB galaxies less clear. Recent observations suggest very strongly that the Galaxy has a small central bar. A photograph of a barred spiral galaxy is shown in fig. 39. The lenticular galaxies resemble spiral galaxies in that they are also highly flattened but they have no spiral structure. Fig. 40 shows a lenticular galaxy.

Figure 37.   A spiral galaxy (Sc). M51 in Canes Venatici. A satellite irregular galaxy
is also shown. (Photograph from the Hale Observatories, Palomar 5 m telescope.)

Irregular galaxies

In addition to the four types of galaxy described above, there are various
types of irregular galaxy. These can to some approximation be divided into two
groups, *IrrI* and *IrrII*, in such a way that IrrI look irregular but are not really very
irregular, whereas IrrII both look and are irregular. The distinction is that in IrrI
galaxies such as our Galaxy's own two principal satellites, the *Large Magellanic*

Figure 38.   A spiral galaxy (Sb) seen edge on. NGC4565 in Coma Berenices. (Photograph from the Hale Observatories, Palomar 5 m telescope.)

Cloud (LMC) and the *Small Magellanic Cloud (SMC)*, the mass distribution is very much less irregular than the light distribution. They share this property with spiral galaxies and IrrI galaxies appear in some sense to be potential spiral galaxies which have not quite realised their potential The genuinely irregular galaxies (IrrII) tend to be either galaxies which have recently suffered a violent explosion or galaxies which are seriously distorted by the close presence of one or more other galaxies. It is, in fact, not easy to see how an isolated irregular galaxy, which is composed of stars with their individual motions, can survive as such for

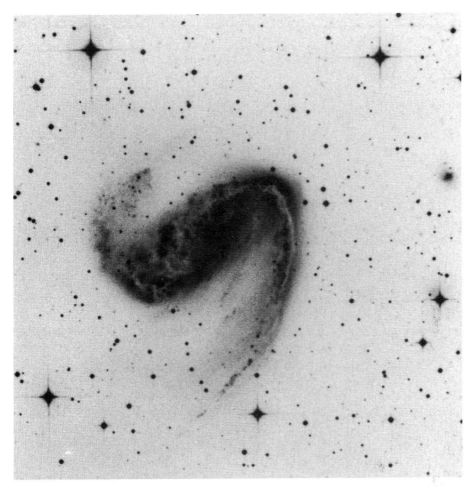

Figure 39. A barred spiral galaxy. NGC2442 in Volans, a negative print. (Photograph taken on the UK 1.2 m Schmidt Telescope, reproduced by permission of the Royal Observatory, Edinburgh.)

thousands of millions of years. It is now believed that most irregular galaxies owe their shape to a recent gravitational interaction with a close neighbour. Such an interaction may pull some material out of the galaxy but it then has a chance to relax to a more regular shape between encounters. The orbits of the two Magellanic Clouds around the Galaxy contain matter such as star clusters and this Magellanic stream may result from close encounters between the Galaxy and the Magellanic Clouds. In addition many galaxies in which violent events are occurring do not have an irregular appearance. I shall return to irregular and peculiar galaxies towards the end of this chapter. Figs. 41 and 42 are photographs of galaxies of types IrrI and IrrII.

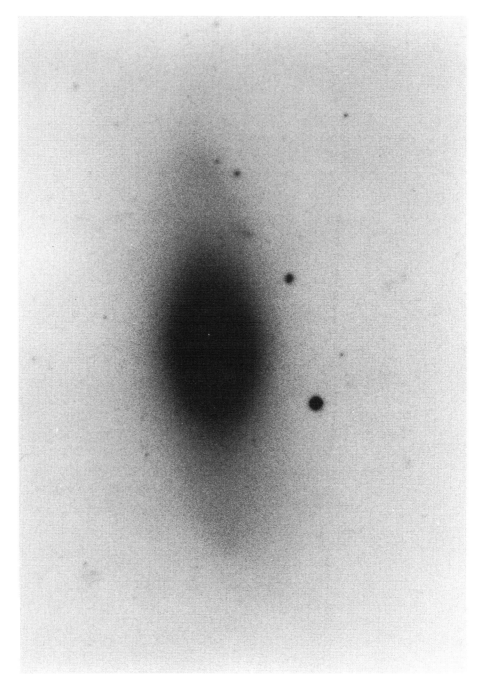

Figure 40. A lenticular galaxy. NGC1380 in the Fornax cluster of galaxies, a negative print. (Photograph taken on the 3.9 m Anglo-Australian Telescope reproduced by permission of the Anglo-Australian Telescope Board.)

Figure 41.   An IrrI galaxy. The Large Magellanic Cloud. (Photograph taken on the UK 1.2 m Schmidt Telescope, reproduced by permission of the Royal Observatory, Edinburgh.)

Possible explanations of the Hubble classification

The Hubble classification described above is essentially the *taxonomy* of galaxies. They have been divided into classes for morphological reasons but at present no physical explanation of the classification has been attempted. As mentioned above, Hubble originally thought that it might represent an evolutionary scheme, so that a galaxy would evolve along the Hubble sequence during its life history. Now this does not appear to be very likely. There are several reasons for this, the strongest of which is perhaps the different distribution of masses between ellipticals and spirals. Both the most massive and the least massive galaxies are ellipticals, with the spirals occupying a much smaller range of masses.

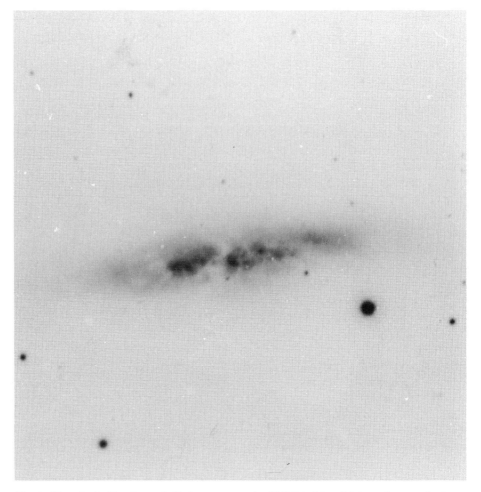

Figure 42.   An *Irr*II galaxy. A *V* electronograph of M82 in Ursa Major taken on the 1 m
telescope of the Wise Observatory, the University of Tel Aviv. (Courtesy D. J. Axon.)

Even the introduction of a different speed of evolution along the sequence for
galaxies of different masses does not seem likely to resolve this difficulty.

It is generally believed today that the difference of morphological type
represents a difference of birth conditions. A cloud of pregalactic gas will have
several properties which can differ from cloud to cloud and these properties will
affect the galaxy which is formed from it. Perhaps the two most important gross
properties are the *mass* and *angular momentum* of the cloud and it is tempting to
ascribe the Hubble sequence for galaxies of the same mass to an increasing
angular momentum per unit mass. Thus the highly flattened galaxies may be
supposed to be those which have the highest angular momentum and by contrast a
non-rotating pregalactic cloud would produce a spherical galaxy. We shall see in
Chapter 8 that the final explanation may not be as simple as this; for example, if

stars form early during the collapse of a pregalactic cloud, the resulting galaxy will be much less flattened than if it completes its collapse as a cloud of gas. In addition a galaxy may form not from a single cloud of pregalactic gas but rather from a merger of several or many such clouds. If the individual clouds have their rotation axes randomly orientated, the most massive galaxies can automatically have the lowest angular momentum per unit mass.

Even if the value of the angular momentum per unit mass is the property which determines whether a galaxy is elliptical or highly flattened, it would appear that variation of a further physical parameter may be required to explain the division of flattened galaxies into spirals, barred spirals, lenticulars and indeed irregulars of the Magellanic type. This additional physical parameter has not at present been identified. Van den Bergh suggests that lenticular galaxies can be classified in a sequence S0a, S0b, S0c parallel to the sequences of spirals and barred spirals, with the size of the nuclear bulge varying along the sequence, so that lenticulars should not be regarded as intermediate between ellipticals and spirals. In addition he suggests that Magellanic (IrrI) type galaxies could be classified IrrIa, b, c. If his suggestion is adopted, it remains necessary to identify the physical parameter which determines (for example) whether a galaxy will be S0a, Sa, SBa or IrrIa.

Although it is unlikely that evolution takes place along the Hubble sequence of galactic type, the division between ellipticals and spirals is probably much less sharp than appears at first sight. There is now strong evidence that the mass distribution in spiral galaxies does not follow the light distribution, in the sense that there is much more mass in the outer region of spiral galaxies than appears from their luminosity. It is usually supposed that spiral galaxies have low density, low luminosity but massive spheroidal haloes which are very much less flattened than the observed spiral. If this is true, the overall difference of angular momentum between spirals and ellipticals of the same mass is much less than appears. There is however a strong suggestion that much of the dark matter in spiral galaxies is not ordinary matter, such as dead stars or extremely low luminosity stars, but is an assembly of weakly interacting elementary particles. If that is the case the ordinary matter in spiral and elliptical galaxies does have a very different amount of angular momentum per unit mass. I shall return to a discussion of *massive haloes* several times later in the book, particularly on pages 124 to 126 in Chapter 5.

There is still not agreement about what relationship, if any, lenticulars have to galaxies of other types. Because S0s are highly flattened, there is a natural tendency to try to associate them with spirals. They are very common in rich clusters where there also tends to be intracluster gas. The suggestion was made that S0s were originally spirals but that in passing through the cluster they had been stripped of their gas, which had been pushed out of the galaxy as a result of collisions with the cluster gas. This explanation was popular for a long time but attempts to show that the idea is quantitatively as well as qualitatively correct have not been successful. More recently there have been suggestions that S0s are more closely related to ellipticals.

Observations of the rotation of spiral galaxies

I have described in the last chapter how the rotation of our Galaxy can be observed. What is the position with regard to other spiral galaxies? To observe the rotation it is necessary to have a galaxy which is edge-on or nearly so, so that there is a reasonable degree of motion towards or away from the observer. In the case of such an edge-on galaxy, it is possible to take a spectrum of the total light output from the galaxy, if the galaxy is neither too near nor too distant, by arranging that the image of the major axis of the galaxy is aligned along the slit of the spectrograph (see fig. 43). Suppose that the galaxy is rotating in such a way that the left end of the major axis in fig. 43 is coming towards us and the right end is retreating. Then, if the galaxy were rotating uniformly, the spectrum would appear as in fig. 44. The spectral lines are tilted because the light from the left side of the galaxy has a Doppler shift to the blue and there is a corresponding shift to the red from the right side of the galaxy; note that these shifts are additional to any shift of the whole spectrum because the galaxy is moving towards or away from us. From such a spectrum it is easy to obtain the apparent rotation velocity of the galaxy and then to deduce the true rotation velocity by making a correction for the inclination of the galaxy, if it is not exactly edge-on. Such a galaxy will have an elliptical outline and the inclination can be deduced from the ratio of the major and minor axes which would be equal if it were face-on.

The spectra that are observed tend to look more like fig. 45 than fig. 44. The shape of the spectral lines in fig. 45 indicates that the galaxy is not rotating uniformly. It is possible from such a spectrum to determine the variation of the circular velocity with distance from the galactic centre and to compare the resulting rotation curve with that found for our galaxy and shown in fig. 28. The main difficulty in doing this is due to the broadening of the observed spectral lines. These are broadened for two main reasons. Because the light of the galaxy is principally the sum of the light from its stars, the spectral lines are broadened because those of individual stars are, mainly because the stars are rotating with one side approaching us and one receding from us. In addition the stars have random motions as well as circular motions. In spiral galaxies, but not elliptical galaxies, these random motions are unimportant compared to the circular motions and it is usually possible to find some spectral lines which are narrow enough for a reliable rotation curve to be produced.

In fact, many rotation curves are obtained by a study of the 21 cm radiation from atomic hydrogen. This can frequently be observed at a greater distance from the centre of the galaxy than reliable information can be obtained from optical spectral lines. When radio observations are used there is one further complication. As I have explained earlier, the dust in the galactic plane absorbs starlight strongly so that it is difficult to see stars except on the nearside of an external galaxy and their light dominates in the spectral line. Radio waves are not absorbed as strongly as light and the observed radiation is made up of emission from all material along the line of sight, as is shown in fig. 46. Although this complicates the interpretations of the observations, it is once again possible for rotation curves to be obtained.

S   S'

Figure 43.   The measurement of a galactic rotation curve. The image of the major axis of the galaxy is aligned along the spectrograph slit SS'.

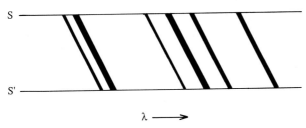

$\lambda \longrightarrow$

Figure 44.   The spectrum of a uniformly rotating galaxy.

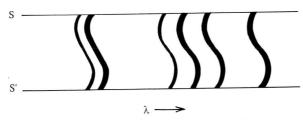

$\lambda \longrightarrow$

Figure 45.   The spectrum of a differentially rotating galaxy.

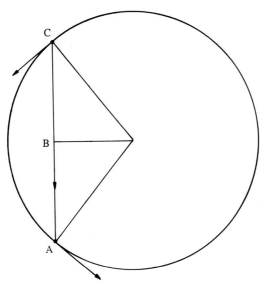

Figure 46.   A spiral galaxy is rotating in the sense shown by the arrows at A and C. An observer viewing the galaxy edge-on receives radiation from the line ABC, which appears as one point on the apparent galactic disk.

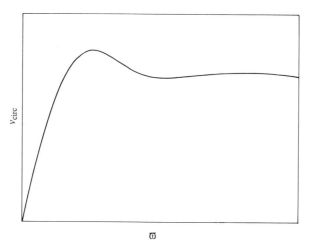

Figure 47.   A schematic galactic rotation curve showing the circular velocity remaining flat at large radial distances.

The observed rotation curves are generally similar to that of our Galaxy with the central regions having a higher angular velocity than the outer regions. As I have already stated in Chapter 2, this property is generally understandable when it is realised that the circular velocity at any radius must be related to the gravitational force towards the centre of the Galaxy by the relation

$$v_{\text{circ}}^2/\tilde{\omega} = -g_{\tilde{\omega}}, \tag{3.2}$$

where $g_{\tilde{\omega}}$ is the gravitational force per unit mass in the $\tilde{\omega}$ direction. Far enough from the centre of the galaxy, the gravitational force will be very similar to that of a point mass so that at large distances we expect to observe

$$v_{\text{circ}} \propto \tilde{\omega}^{-1/2}. \tag{3.3}$$

This is not what is found. For most spiral galaxies the rotation curve is schematically as shown in fig. 47. The circular velocity remains high where there is little luminous matter. This is the strongest evidence for the existence of dark or hidden mass in spiral galaxies. The observed rotation curves can be used to learn something about the mass distribution in, and the total mass of, such galaxies. I shall devote part of Chapter 5 to a discussion of the procedure used and of the uncertainties involved. Meanwhile in this chapter I shall quote the values of masses obtained using this technique. Some actual rotation curves of Sb and Sc galaxies are shown in fig. 48.

It is much more difficult to study the rotation of elliptical galaxies because the random velocities of the stars are comparable with or greater than their circular velocities. In addition the galaxies do not contain clouds of neutral hydrogen. This means that the motions of individual stars must be studied in an attempt to separate a systematic velocity from the random velocities. It is this technique which showed that the rotational velocities were not sufficient to account for the observed flattening of elliptical galaxies.

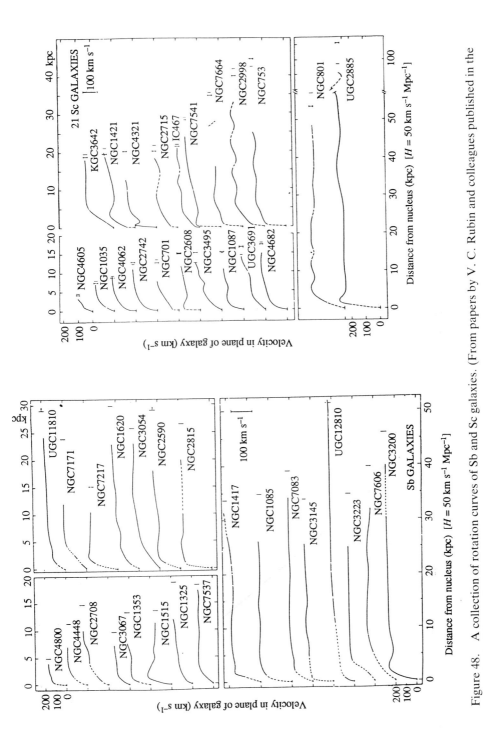

Figure 48. A collection of rotation curves of Sb and Sc galaxies. (From papers by V. C. Rubin and colleagues published in the *Astrophysical Journal*).

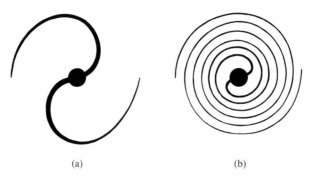

(a)   (b)

Figure 49.   The winding up of a material spiral pattern by differential rotation.

Galactic rotation and spiral structure

Although the *differential rotation* of galaxies can be understood in terms of equation (3.2) in just the same way that there is a very distinct differential rotation of the planets in the solar system described by Kepler's law (equation 3.3), it raises serious problems when we consider the spiral structure of galaxies. Suppose that at some time a galaxy has the somewhat idealised spiral pattern shown in fig. 49(a). Suppose also that the galaxy has a rotation curve of the type which I have described in which the angular velocity near the centre of the galaxy is very much greater than that near to the edge. Now follow the evolution of the galaxy for a time during which the central regions rotate several times but the outer regions do not complete a single rotation. It is easy to see that a structure of the form shown in fig. 49(b) is produced. The spiral structure is rapidly *stretched* and *wound up* and the relatively simple spiral structure is rapidly destroyed. For rotation curves of the type which are observed, very substantial winding up would occur in a time of order 10^9 years compared with an estimated age of most galaxies of order 10^{10} years or more. Because there is a very substantial number of regular spiral galaxies, it appears that the spiral structure is a long-lived phenomenon which must be preserved against the modifications apparently produced by different rotation.

This has led to the idea that the spiral structure cannot contain the same material at all times. If the same gas clouds and stars were always in the spiral pattern, the effects of differential rotation would certainly be those shown in fig. 49. It has instead been suggested that the spiral pattern is a density wave. Where the density of matter is enhanced at the crest of the wave, there is supposed to be enhanced star formation so that both luminous stars and gas show a spiral pattern. In this case, although the spiral pattern is always present, the material in the pattern is continuously changing; this is a similar situation to any other wave such as, for example, a water wave. It is, however, clear that such a density wave must have some complicated properties. It must itself have a differential rotation which counteracts the rotation of the galaxy and which preserves the spiral pattern for much longer than would otherwise be the case. I shall have more to say about spiral structure and the evidence for the wave theory together with other ideas relating to spiral structure on page 144 in Chapter 6.

The properties of galaxies of different types

In the previous chapter I have given a very detailed account of the properties of our Galaxy. I do not wish to repeat such a discussion for any other galaxies. Instead I shall concentrate on a discussion of the similarities and differences in the properties of the various types of galaxy. This is most conveniently done in terms of a small number of *integral properties* of galaxies. Included in these are mass, mass of gas (usually mass of neutral hydrogen), luminosity (in some wavelength range), integrated colour, and chemical composition of the interstellar gas. In this chaper I shall discuss the observed values of these quantities. Later in the book, in Chapters 7 and 8, I shall have something to say about whether a variation in these properties with morphological type can be understood in terms of the processes of galactic formation and stellar evolution in a galaxy.

In principle the simplest of these properties to observe are the luminosity and colour. The apparent luminosity can be observed for any galaxy and this can be converted to absolute luminosity if the distance of the galaxy is known. I have given a brief description of the determination of the distance scale in Chapter 1 and I have made it clear that there are still uncertainties in galactic distances. These will lead to corresponding uncertainties in absolute luminosities of galaxies. The colour of a galaxy is unaffected by its distance unless it is so far away that the red shift of its spectrum is important or unless its light is significantly absorbed by interstellar matter in our Galaxy.

Masses of galaxies

As I have already stated, the determination of galactic masses will be discussed in Chapter 5, but here I shall already make use of the values obtained. The first thing to be said is that the range of galactic masses is very large indeed and the second is that most estimates of galactic masses are likely to be lower estimates. The reason for this, as will be discussed more fully in Chapter 5, is that it is not usually possible to obtain a value for the *total* mass of a galaxy. A value can be obtained for the mass within a certain radial distance from the galactic centre. For a spiral galaxy, this might be the distance out to which the rotation curve can be measured. There may in many cases be a considerable amount of matter in the outer parts of galaxies, which could even increase the estimated mass by as much as an order of magnitude. This must be borne in mind in considering the values given below.

It is impossible to decide which is the lowest galactic mass. There are really two reasons for this. The first is that it is very difficult to discover very small galaxies even if they are very close to our own Galaxy; the second is that we do not have any completely clear definition of what a galaxy is. I can try to define a galaxy as a

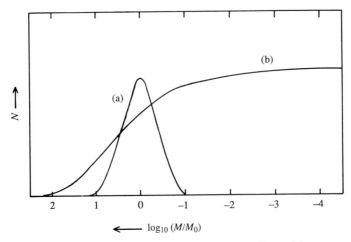

Figure 50.   Schematic distribution of galactic masses. Curve (a) represents an early determination and curve (b) represents more modern data.

large, isolated system of stars. I will not call a globular star cluster a galaxy as it is not sufficiently isolated, but it is not necessarily clear that a rather small system of stars which we may observe is an independent galaxy rather than a globular cluster which has gained independence by escaping from a galaxy. This lack of a precise definition is not really very important and it is known that there are certainly dwarf galaxies which are no more massive than large globular clusters $(10^6 M_\odot)$. There is also uncertainty about the masses of the largest galaxies for the reasons which have been discussed in the last paragraph. There are certainly giant elliptical galaxies with masses greater than $10^{12} M_\odot$ and they may even have masses greater than $10^{13} M_\odot$. This range of masses indicates that our Galaxy, with a mass at least of order $1.5 \times 10^{11} M_\odot$, is above average mass but that it is by no means one of the most massive galaxies.

When galaxies were first studied, it was thought that there was a characteristic galactic mass with a distribution of masses around this value as shown schematically in curve (a) of fig. 50. It is now clear that this distribution is completely false and that it was obtained because of a failure to detect faint low mass galaxies. At present there is no clear evidence of a decline in the number of galaxies at the low mass end and this is indicated by curve (b) of fig. 50. Although the total number of low mass galaxies is very uncertain, it does not appear that they are capable of making any significant difference to the total amount of matter in the form of galaxies. This quantity is very important in a study of the evolution of the whole Universe, as will become clear in Chapter 8. Bearing in mind all that has been said above, we can now display the currently accepted values of galactic masses in Table 5. It can be seen from this table that, as we have already stated, the most massive and least massive galaxies are ellipticals, with the spirals and irregulars occupying a more restricted range of masses.

It has recently become clear that small size is not the only factor which may prevent the discovery of galaxies. Most observed galaxies have rather similar

Table 5: *Galactic Masses. All of the figures
shown are very approximate. If galaxies possess
massive haloes, some masses may be up to ten
times greater than the conventional values.*

Galactic type or name	Mass/$M_\odot$
Massive elliptical	10^{13}
M31	3×10^{11}
Galaxy	$1.5–2 \times 10^{11}$
Small spiral	10^{10}
Typical IrrI	10^{10}
Dwarf elliptical	10^{6}

surface brightnesses. Surface brightness measures the energy an observer receives
from unit area on the sky. It is a galactic property which is independent of the
galaxy's distance from us, because both the apparent size of the galaxy and the
energy reaching the observer fall off as the square of the distance to the galaxy.
Recently some relatively nearby low surface brightness galaxies have been
discovered. If there is a genuinely wide spread in galactic surface brightness, there
could be a large number of galaxies in addition to those which have been
discovered. As such galaxies are not intrinsically small, this could make an
important addition to the amount of matter in galaxies.

The mass-to-luminosity ratios of galaxies

In addition to their masses it is usual to characterise galaxies by the ratios
mass/luminosity (M/L) and (*mass of neutral hydrogen/total mass*)(M_{HI}/M). Not
all galaxies of the same morphological type have the same values of *M/L* and
M_{HI}/M but the scatter in the values for galaxies of any one morphological type is
found to be less important than their variation with galactic type. The variation is
shown in Table 6. Because the values of galactic masses are more uncertain than
luminosities and because these masses may be underestimates, the values of *M/L*
shown may be too small.

It can be seen from Table 6 that *M/L* has the highest value for elliptical galaxies
and the lowest value for Sc galaxies and irregulars and that the opposite is true for
M_{HI}/M. To a first approximation I can say that elliptical galaxies are devoid of
cold interstellar matter, although it has recently become clear that they do contain
some gas and, in some cases, dust. Note that the total mass of interstellar gas may
be significantly greater than M_{HI}. Molecular hydrogen is also present and there
may be as much of it as atomic hydrogen. I shall not discuss the interstellar gas
content of galaxies any further at the moment, apart from making one comment
which looks forward to Chapters 7 and 8. The interstellar gas which is observed in
any galaxy at the present time can have one of three sources. It can be original
galactic gas which has never yet been inside a star, or gas which has been inside a
star and has undergone nuclear reactions before being returned to the interstellar

Table 6. *Mass-to-luminosity ratio and fractional mass of neutral hydrogen as a function of galactic type. All figures are very approximate but show the correct trend. If galaxies have massive haloes the M/L values could be up to ten times higher and the M_{HI}/M values lower. There is, of course additional gas not in the form of* HI.

Galactic type	$(M/M_\odot)/(L/L_\odot)$	M_{HI}/M
E	20–40	~0
S0	10	0.005
Sa	10	0.03
Sb	10	0.05
Sc	<10	0.07
IrrI	<10	0.2

medium, or intergalactic gas which has been accreted by the galaxy in its journey through the intergalactic medium. The mass and chemical composition of the interstellar gas in a galaxy will need to be understood in terms of the processes of star formation and mass loss from stars and possibly of the accretion of intergalactic gas.

The value of *M/L* depends partly on the relative amounts of luminous and non-luminous matter in the galaxy and partly on the distribution of luminosity among stars in the galaxy. In Chapter 2 we have seen that a main sequence star has a luminosity which rises steeply with its mass through a relation which has the very approximate form

$$L_s \propto M_s^4, \tag{2.2}$$

while the total nuclear energy supply of the star satisfies

$$E_N \propto M_s, \tag{2.3}$$

leading to a lifetime of a main sequence star

$$t_{ms} \propto M_s^{-3}. \tag{2.4}$$

We also know that the main sequence lifetime is the main part of the whole lifetime of a star.

Massive stars produce a much higher luminosity per unit mass than low mass stars but they also live a much shorter time. Elliptical galaxies, which have the highest values of *M/L*, also have little or no interstellar gas. This implies that few stars, if any, can be forming in these galaxies at the present time or can have formed in the recent past. Observations confirm that elliptical galaxies do lack massive blue main sequence stars. Because the elliptical galaxies do lack these stars, which are present in spiral and irregular galaxies, it is not surprising that their value of *M/L* is larger than the value for spiral galaxies. It is also reasonable that the integrated colours of spiral galaxies should be bluer than the colours of elliptical galaxies. Although the presence of massive blue stars does affect the value of M/L, it is clear that they cannot be the most important element in any galaxy. All the values of $(M/M_\odot)/(L/L_\odot)$ shown in Table 6 are much greater than unity. This fact that M/L is *large* for all galaxies indicates that the majority of the mass in any galaxy must be contributed either by interstellar gas and dead stars or by luminous stars much less massive than the Sun or by other non-luminous matter such as weakly interacting elementary particles. The last is probably true in most if not all cases. In Chapter 7 we shall see that there is a *tendency* for M/L to increase as a galaxy evolves and that very young galaxies may have had very much lower values of M/L than those which we observe today.

There is one further integrated property of galaxies which I mentioned on page 69. This is the chemical composition of the interstellar gas. Observations of variations of chemical composition from galaxy to galaxy and also from region to region inside galaxies are beginning to become available and these observations should eventually give useful information about galactic evolution. I return to this topic in Chapter 7.

The distribution of galaxies in the Universe: the Local Group of Galaxies

Having summarised briefly the observed properties of galaxies of different types, with the exception of various types of peculiar galaxy which I shall discuss briefly later in the chapter, I now consider the distribution of galaxies in space. Although I have attempted to define galaxies as isolated systems of stars, galaxies are not usually truly isolated systems. Thus, just as a large proportion of stars form double or multiple systems, so also do galaxies. Our own Galaxy has two principal companion galaxies that form with it a gravitationally bound system (the Large and Small Magellanic Clouds) and there are some dwarf elliptical galaxies, which are near to our Galaxy and which are also its companions. New discoveries of such galaxies are made from time to time. Whether or not all multiple systems of galaxies result from single birth processes or whether any of them can have resulted from subsequent capture is perhaps not so clear as in the case of stars. As the ratio of typical galactic size to galactic separation is much greater than the ratio of stellar size to stellar separation, collision or capture is more probable. This is particularly true near the centres of rich clusters of galaxies. For the same reason, there is a greater likelihood that galaxies will be disturbed or distorted by their companions. I have mentioned on page 46 that the outer part of the gas layer in our Galaxy is distorted. Many people believe that this distortion was caused by the gravitational attraction of the Large Magellanic Cloud when it was last at the point in its orbit around our Galaxy which is nearest to the centre of the Galaxy (perigalacticon). It appears that the orbit of the Magellanic Clouds about the Galaxy contains other dwarf galaxies, globular clusters and clouds of gas. It is probable that this *Magellanic stream* is populated

Table 7. *Principal members of the Local Group of Galaxies. M_V is the visual absolute
magnitude (constant $- 2.5 \log L$). The distance from the Galaxy is shown for some of the
more important members. M32 and NGC 205 are satellites of M31. d means dwarf.
Photographs of M31, LMC and NGC 147 can be found on pages 16, 61 and 17*

Name	Type	M_V	Distance/kpc
M31 = NGC224	Sb	-21.1	690
Galaxy	Sb or Sc	-20?	—
M33 = NGC598	Sc	-18.9	690
LMC	IrrI	-18.5	50
SMC	IrrI	-16.8	60
NGC 205	E6	-16.4	
M32 = NGC221	E2	-16.4	
NGC6822	IrrI	-15.7	460
NGC185	dE0	-15.2	
NGC147	dE4	-14.9	
IC1613	IrrI	-14.8	740
Fornax	dE	-13.6	
Sculptor	dE	-11.7	
Leo I	dE	-11.0	
Leo II	dE	-9.4	
Ursa Minor	dE	-8.8	
Draco	dE	-8.6	

with the debris of encounters between the Magellanic Clouds and the Galaxy.
This is somewhat similar to the existence of comets and meteor streams in a single
orbit around the Sun.

As well as galaxies occurring in pairs and triplets, there is a strong tendency for
them to be members of larger groups or clusters, which may be considered in some
sense analogous to galactic and globular star clusters. The Galaxy belongs to what
is called the *Local Group of Galaxies* which has at least twenty members and
which certainly contains other dwarf galaxies whose membership is not yet clearly
established and others which remain to be discovered. The principal members of
the Local Group, with some of their properties, are listed in Table 7. It can be seen
that the Local Group contains a reasonable variety of galactic types but that it
lacks a large elliptical galaxy; this is not surprising when the total number of giant
elliptical galaxies in comparison with galaxies of all types is considered. Even in
the Local Group the Galaxy is not the most massive member, although it and the
Andromeda Galaxy, M31, which is perhaps $1\frac{1}{2}$ times as massive as the Galaxy, but
which is otherwise rather similar, do stand apart from the other galaxies of the
group. The dimensions of the Local Group are also indicated in Table 7 from
which it can be seen that the nearest large galaxy to our own is about 690 kpc away.

Clusters of galaxies

Galaxies are also arranged in groups which are very much larger than the Local Group and which are known as *clusters of galaxies*. A large cluster may contain thousands of galaxies excluding the very faint small members with cannot be detected. The nearest large cluster is known as the Virgo cluster. I have already shown the most massive galaxy in this cluster in fig. 35. The centre of this cluster is believed to be about 15 Mpc from the Galaxy but the Local Group is itself believed to be an outlying satellite of the Virgo cluster. Another very well known cluster somewhat more distant than the Virgo cluster is the Coma cluster in the constellation Coma Berenices. A photograph of this cluster is shown in fig. 51.

Figure 51.   A nearby cluster of galaxies in Coma Berenices. (Photograph from the Hale Observatories, Palomar 5 m Telescope.)

At large distances astronomers tend to study the properties of clusters of galaxies rather than individual galaxies. Just as I have previously classified stars and galaxies, clusters of galaxies are now divided into classes which depend on the number of galaxies which they contain (*richness class*). The most massive and luminous individual galaxies tend to be found in the richest clusters of galaxies. A cluster is first identified because the density of galaxies in a given region of the sky is found to be higher than is typical of the surrounding region. There is then a presumption that the cluster is a gravitationally bound system. This presumption is then strengthened if it is found that the redshifts of galaxies in the cluster are similar and, in fact, those galaxies in the same direction with vastly different redshifts from the average will now be regarded as foreground or background objects rather than as cluster members. Although this identification of clusters of galaxies seems to be very reliable, there are some remaining difficulties which will be discussed further in Chapters 5 and 8. If the clusters are to be gravitationally bound systems, the gravitational pull of all of the galaxies in the cluster must be large enough to prevent individual galaxies from escaping. We can observe the *random velocities* of galaxies in clusters and can deduce what must be the approximate total mass of the cluster to hold it together. This is often found to be much larger than the estimated mass of the cluster unless it contains much mass not in the form of visible galaxies. I shall have more to say about this *missing mass* problem later.

Given that many galaxies do appear to belong to clusters of galaxies, there are various questions which can be asked, such as:

(i) is there any significant number of galaxies which do not belong to clusters?

(ii) do clusters of all possible sizes exist and how important is higher order clustering – clusters of clusters?

There are not at present clear answers to these questions which are discussed by statistical analyses of the distribution of large numbers of galaxies. Superclusters which contain many clusters have however been identified. Considerable progress is likely in this subject in the next few years as modern telescopes and ancillary equipment can detect very much fainter galaxies than in the past and new measuring machines can measure the positions and apparent luminosities of many millions of galaxies.

One further point on which I should comment is the relative proportion of galaxies of different types. In the earliest samples which contained only relatively nearby galaxies and which lacked both dwarf ellipticals and rich clusters of galaxies, the majority of galaxies appeared to be spirals, with a significant number of ellipticals and very few irregulars and lenticulars. It is not easy to give a definite estimate of the relative numbers of different types of galaxy. Ellipticals are more common than spirals in rich clusters but a large fraction of galaxies outside clusters are spirals. The highly flattened galaxies near the centre of rich clusters appear to be lenticulars. In addition the smallest galaxies are dwarf ellipticals which are probably present in very large numbers.

The large scale distribution of galaxies

Now I turn to the distribution and dynamics of galaxies on the largest scale which we can observe. The first observation that can be made is whether or not galaxies occur equally in all regions of the sky, provided only that we are looking at regions significantly larger than those occupied by single clusters of galaxies. In the case of stars in our own Galaxy, we know that the apparently brightest stars *are* spread over the whole sky but that the fainter stars occur preferentially in the direction of the Milky Way, the disk of the Galaxy. A similar result is not obtained for galaxies. In this case there is an approximately *isotropic distribution of galaxies* including the faintest, when allowance has been made for the zone of avoidance which results from our inability to observe galaxies in the direction of the galactic plane. There is no clear evidence of any flattening in the complete system of galaxies which we observe, although many individual clusters of galaxies are certainly not spherical.

It has become apparent in the 1980s that the distribution of galaxies possesses much more structure than was previously thought. There are regions in space, voids, which contain essentially no galaxies and which can have dimensions measured in tens of Mpc. In addition there appear to be long chains of galaxies. It is believed that the large scale distribution of galaxies is telling us something about the problem of galaxy formation, which I shall discuss briefly in Chapter 8. Early studies of the galaxy distribution by Hubble gave a much more uniform distribution than is observed today; his studies were of relatively nearby clusters of galaxies. These observations led to the development of models of a homogeneous, isotropic expanding universe, which I shall describe shortly. Today it is, in contrast, observations of the cosmic microwave radiation to be discussed in Chapter 8, which lead to the belief that the Universe was once very uniform. A major problem is then to understand how the very smooth early Universe became the very structured Universe which is observed today. The isotropy mentioned in the last paragraph is an average of a much more irregular distribution.

The expansion of the Universe

The second important observation is one which I have already discussed in Chapter 1. The spectral lines from distant galaxies are shifted to the red and the redshift is related to the distance of the galaxies from us. This was established by Hubble in the 1920s, although the first results had been obtained somewhat earlier. If this redshift is interpreted as a Doppler shift, which is the interpretation which we have used in discussing the motions of stars in galaxies and the motions of galaxies in clusters, this means that there is a general expansion of the Universe, with the most distant galaxies moving away from us more rapidly than the nearby ones. Although there have from time to time been attempts to suggest alternative causes of the redshift, most astronomers are agreed that the most plausible explanation of the redshift in the spectra of distant galaxies is one of universal expansion.

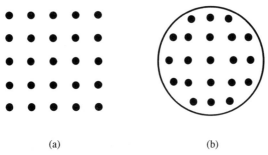

(a)   (b)

Figure 52.   A homogeneous distribution of objects in an open (a) or closed (b) expanding Universe.

Is the Galaxy in a special position in the Universe?

These two observations of an isotropic expanding system of galaxies, as seen from the Earth, might appear to place our Galaxy in a special position in the centre of the Universe. If the Universe possesses a unique centre that is certainly necessary, but it is quite easy to see that, either if the Universe is infinite in spatial extent or if its space is curved, it is possible for all points equally to appear to be at the centre. Two such examples, which are certainly not applicable to the real Universe, are shown in fig. 52. In fig. 52(a) dots are evenly spaced throughout an infinite two-dimensional Euclidean space, while in fig. 52(b) they are supposed to be uniformly spread over the surface of a sphere. In either case if the distances between all the dots are increased uniformly, in the latter case by inflating the sphere, each dot will see all of the others running away from it with a velocity proportional to distance. The general theory of relativity *does* suggest that the Universe has a curved space-time and it does admit models of the Universe which are *homogeneous, isotropic* and *expanding*. Homogeneous means that the properties of the Univere are on the average the same at all points and in particular there is not a unique centre of the expansion. I shall discuss these models of the expanding Universe further in Chapter 8.

Peculiar galaxies and quasars

So far in this chapter I have confined my discussion of the properties of galaxies to what might be called normal galaxies. I conclude the chapter by saying something about a minority of galaxies which have unusual properties and I also discuss whether this means that some galaxies are permanently normal or whether most galaxies may have periods when they appear peculiar. I start with a category of object which is perhaps the most controversial, the quasar.

Quasars

The history of the discovery of quasars is very interesting. When radio astronomy was developed it was discovered that there were point sources of radio waves in the Universe in just the same way as stars appear to be point sources of light. These objects were given the name *radio stars*. The spatial resolution of the earliest radio telescopes was very poor but, as the resolution was improved, it was discovered that most of these radio stars were extended sources of radio emission and that many of them could be identified either with gas clouds in our Galaxy or with other galaxies.† The name radio star lingered on for a long time after it was doubtful whether there were real radio stars but eventually the name radio source was adopted and this is in general use today. Then in 1962 four radio sources were discovered whose positions appeared to be identical with what were believed to be four faint stars. It was believed that the first true radio stars had been discovered but that belief did not last very long. Originally it proved impossible to understand the spectra of these *stars* but then it was realised that the spectra could be interpreted if they had redshifts comparable with, or even much greater than, those of the most distant galaxies which had been observed.

Because the objects looked like stars they were given the name *quasi-stellar objects* which was abbreviated to *quasars*. Although quasars are believed to be very much larger than stars, they can still appear as point sources of light if they are at the very large distance implied by the cosmological Doppler interpretation of their redshifts. They must, however, be very much smaller than ordinary galaxies. If the redshift is caused by cosmological expansion quasars are the most distant objects known in the Universe and their light output is greater than any other known object, ranging up to 10^{41} W. Although the first quasars were discovered because of their radio emission, it is their optical properties which are the most remarkable. By this I refer particularly to a luminosity very much higher than an ordinary galaxy from a volume much smaller than a galactic volume. As soon as it became possible to discover quasars without reference to their radio properties, many radio-quiet quasars were found, and radio astronomy plays no rôle in most new discoveries of quasars. In recent years the record for the highest quasar redshift has frequently been broken and it now stands at 4.9. The light from the quasar probably set out when the Universe was less than ten per cent of its present age.

Quasar absorption lines

A minority of astronomers does not accept that the redshifts of quasars are Doppler shifts of cosmological origin and I shall comment on this later. One further property of quasars should, however, be mentioned immediately. The redshifts of quasars were originally found from emission lines (fig. 53) rather than from absorption lines (fig. 54) which are more often used in the study of stellar spectra. However, in some quasars both absorption lines and emission lines have been identified and in these cases some remarkable properties have been found.

---

† There *is* intermittent radio emission from the Sun, particularly at a time when there are many sunspots, but such emission would be too weak to be detected if the Sun were as far away as a typical star. Radio emission from some other stars, which are much more powerful emitters than the Sun, has now been detected.

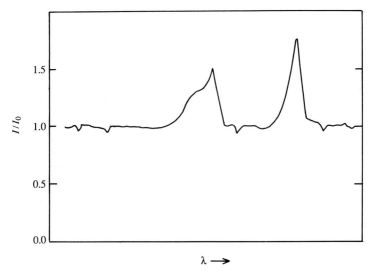

Figure 53.   Two strong emission lines in a quasar spectrum.

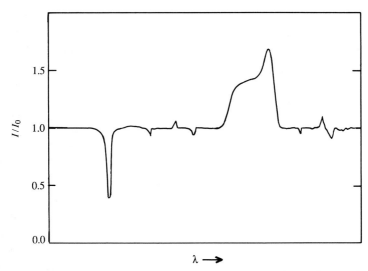

Figure 54.   An absorption line in a quasar spectrum is shown displaced to the blue of a corresponding emission line.

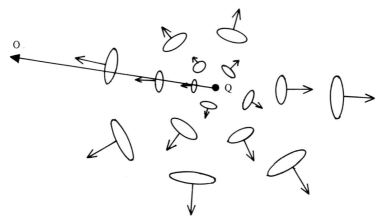

Figure 55.   A possible arrangement of clouds moving radially outwards in the envelope of a quasar Q, showing several in the line of sight of an observer O.

In some quasars there are several sets of absorption lines with different redshifts which in turn differ from the emission redshift. If these redshifts are all caused by the Doppler effect they indicate relative velocities between the absorbing and emitting regions which are a significant fraction of the speed of light. Either they arise from large relative velocities in the quasar itself or they arise from the presence of intergalactic gas clouds between the quasar and the observer, which absorb the quasar's light. In the former case gas clouds may be being expelled with high velocities from the central regions of the quasar (fig. 55). The latter explanation is precisely similar to the process of absorption of starlight by interstellar clouds which was discussed in Chapter 2. The large relative redshift would then arise if the intergalactic clouds were sufficiently far from the quasar that they possessed very different cosmological expansion speeds. Unfortunately the distribution of quasars on the sky is not sufficiently dense that the intergalactic cloud hypothesis can be verified by noting the same Doppler shift in the spectra of neighbouring quasars. Although the origin of the absorption redshifts is not completely clear it now seems likely that in most cases they are of external origin but that there are also probably some cases in which the redshifts arise from local motions near the quasar. This is almost certainly true in those rare cases where there is an absorption redshift greater than the emission redshift.

The strongest evidence for intermediate clouds comes from those quasar spectra which show very large numbers of absorption lines which are attributed to the Lyman α spectral line of atomic hydrogen. This Lyman α *forest* suggests that there are many gas clouds beween the quasar and the observer. They could be true intergalactic clouds or they could be gas in disks of intermediate spiral galaxies. The number of absorption lines correlates with the redshift of the quasar, which is itself strong evidence for the intermediate cloud hypothesis. There is now great interest in trying to discover other spectral lines than those of hydrogen as this would give some information about the chemical composition of the clouds. Are

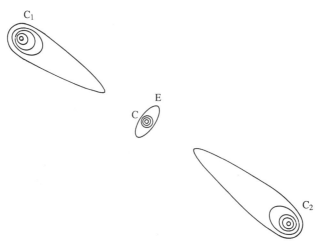

Figure 56.   A double radio source. Components C_1 and C_2 are placed on either side of an elliptical galaxy E. A central radio source C is also shown. The curves are contours of equal radio brightness.

they truly primeval intergalactic clouds which have never been involved in star formation or are they galactic clouds including material that has been expelled from stars?

Another exciting discovery involving quasars was that of gravitational lenses. One of the key predictions of the general theory of relativity was that light rays could be bent by passing close to massive objects. Furthermore in some circumstances two or more images of the source of the light could be produced by a *gravitational lens*. In 1979 two quasars were discovered very close together on the sky with very similar redshifts and luminosities and essentially identical spectra. It was immediately suggested that what was being observed was two images of one quasar produced by an intermediate galaxy. Since then more such gravitationally lensed images have been detected and they are potentially an important probe of matter along the line of sight. More recently continuous luminous arcs have been interpreted as images of single objects stretched across the sky by the gravitational lens effect. Note that the travel time to the observer of the light producing different images will not be the same and, if the quasar has a variable light output or spectrum, there will be a resulting time delay in the variability.

Radio galaxies

I now leave quasars temporarily and consider some other objects which are definitely galaxies. I have already mentioned that some radio sources can be identified with galaxies. All galaxies have some radio emission which arises from a variety of processes within them but a subclass of galaxies are very intense emitters of radio waves and have been given the name *radio galaxies*. The simplest type of radio galaxy is the *double radio source* (fig. 56). In this case we have a

Figure 57.   The helical motion of a charged particle about a line of magnetic induction.

galaxy (usually a very large elliptical galaxy) and on either side of it but some considerable distance away from it there are regions of intense radio emission. In addition there is often weak radio emission between the main double sources and the galaxy and there may also be a further strong and very compact source centred on the galaxy itself. As observations accumulate, it appears quite possible that there is always a strong central source. Although we have no method of measuring the distance to the regions of radio emission, the manner in which they are placed relative to the optical galaxy makes it difficult to believe that there is not a real connection between them and the galaxy. Not all radio galaxies have a structure which is as simple and clear cut as this; some have more than two radio components and in others the relation to an optical galaxy cannot be established, presumably because it is too distant to be visible. However, the existence of radio galaxies of this type is quite clear.

The energy content of radio sources

It is generally believed that a radio galaxy of the above type arises as a result of an explosion or some other violent activity in the centre of the galaxy. The amount of energy involved in such an explosion must be very large. The most plausible explanation of the radio emission from these and most other types of radio source is that it is produced by charged particles moving in a magnetic field (fig. 57). Charged particles move in a helical path in a magnetic field and because their motion is accelerated they radiate energy. For appropriate values of the initial energy of the particle (electron) and of the strength of the magnetic field, this radiation is produced in the radio region of the spectrum. For electrons with energy $\mathscr{E}$ in a magnetic induction **B**, the main radiation is produced in the neighbourhood of the frequency

$$v_{max} = \frac{eB_\perp}{m_e} \left(\frac{\mathscr{E}}{m_e c^2} \right)^2, \tag{3.4}$$

where $B_\perp$ is the component of the magnetic induction perpendicular to the direction of motion of the particle. The radiation is known as *synchrotron radiation* because the same effect produced radiation in the particle accelerator known as the synchrotron. The intensity of the radiation depends on both the total energy possessed by the particles and on the magnitude of the perpendicular component of **B** according to the formula

Table 8. *Minimum energy requirements for three extragalactic radio sources*

Name of source	Energy/J
Cygnus A	2×10^{52}
Centaurus A	10^{53}
3C 236	10^{54}

$$P_{rad} \approx 1.6 \times 10^{-14} B_\perp^2 \; \mathscr{E}\mathscr{E}_{tot}/m_e^2 c^4 \; \text{W},† \qquad (3.5)$$

where P_{rad} is the radiated power and $\mathscr{E}_{tot}$ is the total particle energy. A given radiated power can be produced either by a large particle energy in a weak magnetic field or by a lower particle energy in a stronger field, but it is possible to show that a given intensity of radiation implies a minimum total energy present in the form of charged particles *and* magnetic fields. This minimum energy can be extremely large (Table 8).

Even this estimate must be substantially increased when it is realised that there may be other matter present in the radio source components which does not contribute to the observed spectrum; this may include both high energy nuclei and cold matter. The nuclei *must* be present to balance the electric charge of the electrons and they *may* increase the energy content by one or two orders of magnitude. Even though there may also in some circumstances be other mechanisms of energy loss which are more effective than synchrotron radiation, the total energy content of the radio components of radio galaxies must be very large. I cannot in a book of this length attempt to discuss theories of radio source structure, which are themselves very controversial. It is, however, possible to say that there is now general agreement that the energy was not all deposited in the components in an initial explosion and that the energy travels to the components either in the form of electromagnetic waves or as relativistic charged particles. The most intense emission arises where the beam impacts the intergalactic medium. This theory implies intense activity in the central regions of the galaxy and suggests some similarity with quasars. In addition the radio emission of some quasars has been found to have a double structure similar to but on a smaller length scale than that of radio galaxies.

Other active galaxies – Seyfert galaxies

Earlier I introduced the expression 'peculiar galaxy'. Peculiar galaxies can be divided into two types. There are those which are sufficiently close to another galaxy that they are seriously distorted by the gravitational force of the other galaxy. I shall not discuss these further here. There is in addition the class of active galaxies which are undergoing some type of violent event of internal origin.

---

† Here it has been assumed that all electrons have the same energy $\mathscr{E}$. In reality they will have a spectrum of energies. The equation will still be correct with $\mathscr{E}$ as an appropriate mean energy.

The radio galaxies belong to this class. There are other types of active galaxies which may radiate strongly in the radio and X-ray regions of the spectrum but which were first discovered because their optical spectra include very broad emission lines. The structure of these emission lines can be understood if it is assumed that gas is flowing outwards from the central regions of the galaxy following an explosion. One such class of emission line galaxies is known as *Seyfert galaxies* after their discoverer K. Seyfert; unlike the strong radio galaxies many Seyfert galaxies are spirals. A related group of galaxies are known as BL Lac objects. It has become obvious in the recent past that explosive events are very common in galactic nuclei. The intensity of the explosive activity varies very greatly from galaxy to galaxy. Even in our own Galaxy there is evidence for recent violent activity which has led to the outward motion of gas which I have mentioned in Chapter 2. The aftermath of a single violent event cannot last for ever and it is not believed that active galaxies are generically different from the normal galaxies which I have discussed in the Hubble classification at the beginning of this chapter. Rather, it is thought that any galaxy may from time to time become active. The form and degree of activity may depend on Hubble type and on the age of the galaxy and, possibly, on whether or not the galaxy is in a cluster.

The energy source for violent activity

I have explained that a very large energy release is required to power the double radio sources. This is also true of quasars if they are at cosmological distances. Because it proved difficult to understand how the energy release of quasars could be achieved, some researchers suggested that quasars were much nearer to us and hence less luminous. In that case the redshifts either are caused by very high velocities which are not cosmological or have a non-Doppler (and possibly unknown) origin. In either case much remains to be explained. Attempts to discover the distance of quasars have concentrated on trying to demonstrate that quasars with low redshifts are associated with clusters of galaxies with similar redshifts or to show that quasars are physically connected with galaxies with distinctly smaller redshifts. The first result would support the cosmological interpretation whereas the second would imply that some part of the redshift was non-cosmological. Claims of both types of observation have been made and disputed but there is now a variety of reasons for believing that quasars are extremely luminous and at very large distances. Many quasars are surrounded by a fuzz which suggests that they are inside a galaxy or at least a forming galaxy. In fact there is increasing evidence that all quasars which are near enough for a surrounding galaxy to be seen are inside galaxies. Others appear to be associated with galaxies with a similar redshift. Finally there are the properties of the Lyman α forest already mentioned.

Whatever is the status of the quasars, the need for a very large rate of energy release cannot be evaded in the radio galaxies. Originally it was supposed that the energy for a galactic explosion would be thermonuclear in nature; nuclear reactions in a concentrated mass in a galactic nucleus would release a large

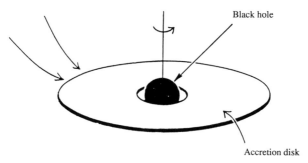

Figure 58. An accretion disk around a black hole. Matter falling towards the black hole, as in the direction of the arrows at the left, forms a rotating flat disk. As a result of viscosity in the disk, it gradually moves towards the centre. No matter can have a stable orbit in the gap shown between the accretion disk and the black hole.

amount of energy. Then it was realised that, as nuclear reactions in a mass M cannot possibly release more than just under one per cent of the rest mass energy Mc^2, some of the radio galaxies would require very large masses to explode. In extreme cases we might be talking about the explosion of $10^9 M_\odot$. Although there is at present no complete explanation of the source of the energy, it is now believed that gravitational energy release must also be important. According to the general theory of relativity, there is a maximum possible release of energy equal to the rest mass energy if a body collapses to a point. It seems clear from detailed discussions that this limit cannot be attained but a release of $0.1Mc^2$ might well be possible before an object falls within its *Schwarzschild radius*†

$$R_{Sch} = 2GM/c^2, \tag{3.6}$$

from which no further energy can escape, and becomes what is known as a *black hole*. Models usually assume that matter is accreted to an existing black hole. If particles fall in radially with no collisions with other particles, no useful energy release will occur. As accreted particles can be expected to have non-zero angular momentum about the central object, they will tend to form an *accretion disk* around it (fig. 58) and it is the energy dissipated by friction in such a disk which leads to the useful energy release. It seems clear that it will be some time before the properties of active galaxies are fully understood. There are indeed still some suggestions that the energy released in at least some of them is provided by the luminosity of a very massive and compact star cluster.

There is some evidence for the existence of massive compact objects in the nuclei of galaxies. If quasars are the nuclei around which some galaxies form,

---

† The probable existence of this critical radius can be understood without consideration of the general theory of relativity. According to Newtonian mechanics the escape velocity from a body of mass M and radius R satisfies $v_{esc}^2 = 2GM/R$. According to the special theory of relativity the speed of energy propagation by any method must be less than the speed of light. A combination of these two results leads to the prediction of the critical radius (3.6).

there should be dead quasars in the nuclei of the galaxies once insufficient gas remains to fuel quasar activity. The same should be true, but with smaller masses involved, in the nuclei of other galaxies, which may or may not still be active. The presence of massive compact objects in the nuclei of galaxies such as NGC4151 and M87 has been demonstrated by the speed with which stars orbit in the central regions. Although the existence of massive objects is not disputed there is still not agreement about whether they must be black holes or whether they might be compact star clusters.

Summary of Chapter 3

Galaxies can be classified according to their visual appearance. A majority of them have a rather regular appearance and these are in turn divided into ellipticals, lenticulars, spirals and barred spirals. The ellipticals, many of which are triaxial, have three axes of comparable size, but the other three classes are highly flattened. In addition to the regularly shaped galaxies there are two groups of irregulars; of these IrrIs have an irregular light distribution but a much more regular mass distribution, while the IrrIIs are really irregular. In the highly flattened galaxies ordered motion of galactic rotation is much more rapid than the random motion of the stars but rotation is much less important in ellipticals. Spiral galaxies are observed to rotate non-uniformly and this means that a long-lived spiral structure cannot be a material structure but must be a wave pattern.

The range of galactic masses is very large with giant ellipticals possibly having masses as large as $10^{13} M_\odot$ and dwarf ellipticals masses of $10^6 M_\odot$ or less. The masses of spirals and most irregulars are intermediate between these extremes. All galactic masses are very uncertain and many galaxies could possess large masses in extensive haloes. Values of *mass/luminosity* and *gas mass/total mass* vary significantly with galactic type, with spirals and irregulars having both more gas and lower values of M/L than ellipticals. The lower values of M/L are partly caused by the presence of massive bright blue stars in spirals and irregulars but all values of M/L are so large that most of the mass of all galaxies must be in the form of low mass luminous stars or dead stars or weakly interacting elementary particles.

Many galaxies belong to double or multiple systems and they may also be members of larger groups or clusters. Our Galaxy has two principal companions, the Magellanic Clouds, and it also belongs to the Local Group of Galaxies of which it is the second most massive member. The Local Group is also a satellite of a large cluster of galaxies known as the Virgo cluster. It is still not completely clear whether almost all galaxies belong to clusters and what is the extent of higher order clustering. On the largest scale the distribution of galaxies is approximately isotropic and distant galaxies have large redshifts in their spectra which imply that the Universe is expanding, if the Doppler interpretation of the redshift is correct. In detail the distribution of galaxies is far from smooth. There are large voids with dimensions of tens of Mpc, with few if any galaxies, and long linear chains of galaxies.

A minority of galaxies, which may be classified as irregular or may basically be spirals or ellipticals, are described as peculiar or active. These include several types of galaxy exhibiting violent events in their nuclei. One group consists of radio galaxies, which have extensive regions of radio emission on either side of the parent galaxy. Another group contains the Seyfert galaxies in which gas appears to be moving outwards from the nucleus following an explosion. Many galaxies probably have periods in which they are peculiar separated by longer, more normal, phases. Quasars, which have the highest redshifts known and which are very compact objects with optical luminosities higher than galaxies if their redshifts are cosmological, appear to be closely related to galactic nuclei. The energy source in the quasars and active galaxies is not fully understood but it may involve the release of gravitational energy by matter falling into a black hole.

4

Stellar dynamics†

Introduction

In this chapter I shall consider that a galaxy consists of stars alone. To a first approximation this *is* true of a real galaxy apart from the possibility of *hidden mass* in the form of weakly interacting elementary particles; in what follows the gravitational field of any hidden mass is assumed to be added to that of the stars. For most of the life history of a galaxy the mass in the form of stars is much greater than the mass of the interstellar gas and the stars exert a much stronger gravitational influence on the gas than the reverse. Although there is a continual mass exchange between the stars and the gas, this occurs slowly compared to the time taken by individual stars to travel about the galaxy, certainly once the formation phase of the galaxy is completed. To a large extent in what follows I shall refer to the Galaxy, but it may be regarded as typical of other galaxies.

The first point to be made about the Galaxy, if it is considered as a system of stars, is that to a good approximation stars may be regarded as point masses. Except in the densest regions of galaxies, stellar separations very much smaller than 10^{16} m are rare, whilst only the very largest stars have radii greater than 10^{10} m. Thus the ratio of stellar radius to stellar separation is usually less than 10^{-6} and often very much less than this; in the case of the Sun and the nearest known star to the Sun it is 2×10^{-8}. This may be compared with a typical value for gas molecules in air at s.t.p. which is of order 1/50. This indicates that, when I am not considering the detailed physical properties of individual stars, it makes sense to treat the system of stars as a gas and to make use of concepts that are familiar in the *kinetic theory of gases*. It is of course true that many, if not most, stars are members of binary or multiple systems where the stellar separation may be a very much

† The subject matter of this chapter is somewhat different from most of the material covered in the remainder of the book and it also contains more advanced mathematics than the other chapters. Its detailed arguments can be omitted at a first reading without causing difficulty in understanding the remainder of the book. The points from the chapter that are used elsewhere are mentioned in the summary at the end of the chapter.

smaller multiple of the stellar radius. In this case I can regard the binary or multiple system as a gas molecule and I do not need to change my approach.

A galaxy consists of a very large number of stars (at least 10^{11} in a medium or large sized galaxy) which influence each other through their gravitational interaction. Although the gravitational forces between the stars are well understood, it is clearly absolutely impossible to study the motions of all the stars and, even if we could, we should obtain a lot of detailed information which we do not want. For that reason I shall generally adopt a statistical or kinetic theory approach. I shall, however, show later in the chapter that some interesting information can be obtained by studying the motion of one star in the average gravitational field produced by all of the others. This is known as a *test particle* approach and it can be used, for example, to find how large a fraction of the volume of the Galaxy has been explored by the Sun during its life history.

Velocity distribution functions

In the kinetic theory of gases it is usual to discuss a system in terms of what is called a *velocity distribution function*. In a gas which contains only one type of particle (for example molecular hydrogen) it is usual to define a distribution function $F(x,y,z,v_x,v_y,v_z,t)$, where the position of any molecule is (x,y,z) and its velocity components are (v_x,v_y,v_z), in such a way that the number of molecules in the volume element $\delta x \delta y \delta z$ centred on (x,y,z) and with velocities in the volume element in *velocity space*, $\delta v_x \delta v_y \delta v_z$, centred on (v_x,v_y,v_z), at time t is

$$\delta N = F(x,y,z,v_x,v_y,v_z,t)\delta x \delta y \delta z \delta v_x \delta v_y \delta v_z. \tag{4.1}$$

In the case of a stellar system a somewhat different definition is used. Because stars have different masses and because all values of mass are possible, it does not make sense to define a distribution function in terms of the number of stars. Instead it is more useful to define a *mass distribution function* which is such that

$$\delta M = f(x,y,z,v_x,v_y,v_z,t)\delta x \delta y \delta z \delta v_x \delta v_y \delta v_z \tag{4.2}$$

is the total mass in the form of stars in the volume element in position and velocity space. This makes sense because the gravitational force per unit mass exerted by all the other stars is independent of the mass of an individual star.

The Maxwellian distribution

In most elementary problems in the kinetic theory of gases, particularly those in which the molecules are electrically neutral, we are used to the distribution function F departing only very slightly from a specific form, the *Maxwellian distribution*

$$F = n(m/2\pi kT)^{3/2}\exp(-m(v_x^2 + v_y^2 + v_z^2)/2kT), \tag{4.3}$$

where n is the number of particles per unit volume and T is the temperature of the gas. If we study the behaviour of a gas in a box with a velocity distribution which is initially non-Maxwellian, it is possible to show that the effect of collisions of gas molecules with one another and with the walls of the box will be such as to cause the distribution function to approach the Maxwellian form very quickly. Usually the time taken is short compared to the time available for an experiment and, as a result, consideration tends to be restricted to distribution functions which depart only slightly from the Maxwellian form. One particular property of the Maxwellian distribution is that the distribution of velocities in all three directions is the same.

The situation is quite different when we consider stars in a galaxy. In the first place, from observations of stars in the solar neighbourhood in our Galaxy, it is clear that the distribution of velocities does not have a Maxwellian form. As we have seen on page 28 in Chapter 2, when we have subtracted out the velocity of galactic rotation, the mean random velocities of the stars are not the same in all directions as they would be if their distribution had the form (4.3). In the second place there are good theoretical reasons, which I will now describe, for not being surprised that the distribution is not Maxwellian.

Stellar collisions

We have already seen that the stellar gas in the Galaxy is a very diffuse gas. One consequence of the ratio of stellar separation to stellar size being very large is that collisions between stars are very infrequent. They are, in fact, so infrequent that, as I shall show shortly, a typical star in our Galaxy has not collided with another star since the Galaxy was formed. In a normal gas it is the collisions between the molecules which transform an initially arbitrary distribution into the Maxwellian form. If there have been virtually no collisions in a stellar gas, there is no obvious reason why the distribution should be Maxwellian.

Of course, when I discuss collisions between stars or gas molecules, I do not mean collisions in quite the same sense as those between two billiard balls. A close passage of two stars will be called a collision† if the result of their gravitational attraction is to change their direction of motion through a significant angle of order a right angle. It is possible to make a very crude estimate of how close two stars of equal mass must approach one another for this to happen. The effect of their mutual gravitational force will be really important if, at their distance of closest approach, their mutual gravitational potential energy is greater than their kinetic energy of relative motion. If I consider a co-ordinate system in which one star is at rest and the other has velocity v and the distance of closest approach is d (fig. 59), this approximate condition can be written

$$Gm^2/d > mv^2/2 \qquad (4.4)$$

where m is the mass of either star. Inequality (4.4) can alternatively be written

$$d < 2Gm/v^2. \qquad (4.5)$$

---

† In very dense systems, when genuine stellar collisions are probable, those collisions must be given another name such as encounter.

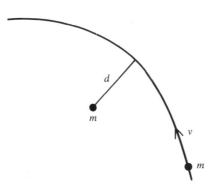

Figure 59.   The motion of a star relative to another at rest.

If I consider a star moving through the galaxy with a velocity v relative to the centre of mass of the other stars, I can regard it as having an effective radius $2Gm/v^2$ for collisions. What I mean by this is that I can regard it as colliding with all stars contained in a cylinder of radius $2Gm/v^2$ about its path (fig. 60). This expression is approximate for several reasons; (4.4) was itself approximate, not all of the stars have the same mass, the star will possibly be interacting significantly with more than one other star at a time, because gravitation is a long range force, and the other stars have random velocities which I have neglected. However, this approximate discussion is sufficiently good for our present purposes. If the number density of stars is n, it is possible to see that the *mean free path* between collisions, which has the same significance as the mean free path, λ, in an ordinary gas, is

$$l = 1/\pi(2Gm/v^2)^2n \tag{4.6}$$

and that the *mean free time* between collisions, τ_c, is

$$\tau_c = v^3/4\pi G^2 m^2 n. \tag{4.7}$$

In fact, as I have said, all stars do not have the same mass and a single star does not move through a background of other stars all of which are at rest. However, if one uses mean values of m and v, the result obtained should be approximately correct.

I can obtain a crude idea of the value of τ_c by using values of v, m and n appropriate to the Sun and the solar neighbourhood. The Sun's velocity relative to the centre of mass of other stars is about 20 km s^{-1}, its mass is 2×10^{30} kg and the nearest known star is rather more than one parsec distant. If I use the solar v and m and use a density of 1 pc^{-3}, τ_c is 3×10^{13} years. This is considerably longer than the believed age of the Galaxy (1–2 $\times$ 10^{10} years). Although the discussion given here is very simplified and is certainly wrong in detail, it is not possible that the value of τ_c has been overestimated by three powers of ten. The discussion is therefore sufficient to show that the gas of stars in the Galaxy is a somewhat unusual gas; it is a *collisionless* gas. As the Maxwellian velocity distribution, which we normally

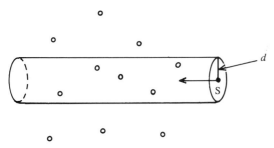

Figure 60.   Collision cross-section. A star behaves as if it has a cross-sectional area πd^2 (where $d = 2Gm/v^2$) and collides with any other star in the cylinder shown. The number of such stars is greatly exaggerated in the figure.

expect, is produced by collisions, we should not be surprised that the stellar gas does not have a Maxwellian distribution.

However, although the stellar velocity distribution does not appear to be Maxwellian, it *does* appear to be essentially time-independent in most galaxies. In saying this I am making an assertion about something which cannot be observed, because no significant change would occur in less than 10^7 or 10^8 years. The reason for the assertion is that most, but not all, galaxies have rather regular shapes. If the velocity distributions were not approximately time-independent, it would be a coincidence for so many galaxies to look symmetrical at the same time. More precisely, the time taken for a typical star to cross a galaxy is likely to be no more than a few times 10^8 years whereas the galaxy probably has an age of over 10^{10} years. This means that there has been ample time for any irregularity in the velocity distribution to become apparent. One *might* perhaps suppose that all galaxies that did not have a sufficiently regular distribution of stellar velocities have dispersed or collapsed long ago and that only the regular ones have survived. Such an argument based on chance alone would seem to indicate far too small a number of surviving galaxies. In fact, it appears that, although collisions are unimportant once stars have formed, the long range gravitational interaction during galaxy formation, when most of the material was gaseous rather than stellar, must have been effective in producing both a regular galactic shape and a regular initial velocity distribution in most galaxies. At the present time in our Galaxy, stars are still being formed out of interstellar clouds and the velocity distribution of newly formed stars must be determined by the velocities of, and within, the gas clouds out of which they are formed. Collisions between interstellar clouds are much more frequent than collisions between stars because clouds have a much lower density than stars and are therefore relatively and absolutely very much larger. Collisions between gas clouds may establish a regular distribution of velocities which will be reflected in the velocity distribution of the young stars. Stars are also much more likely to collide with gas clouds than with other stars and it is believed that the higher random velocities of old stars in the galactic disk, which have been mentioned in Chapter 2, have arisen at least partially from such collisions.

Equation for the distribution function

I will now accept the observation that the velocity distribution functions are time-independent in many galaxies and ask what forms such distribution functions (f) may have. Before I can do this I need an equation for f, which will tell us how an arbitrary distribution function will evolve with time and which will hence be able to tell us what form the distribution function must have if it is not to change with time. Since collisions between individual stars are unimportant except possibly in the dense nuclear regions of galaxies and in star clusters, I will suppose that the gravitational field in the galaxy can be regarded as a smoothly varying function of position and that the stars move in this smooth field without collisions. If the smooth gravitational potential is Φ, an individual star will have the equations of motion

$$\left.\begin{aligned} \dot{v}_x &= \ddot{x} = \partial\Phi/\partial x, \\ \dot{v}_y &= \ddot{y} = \partial\Phi/\partial y, \\ \dot{v}_z &= \ddot{z} = \partial\Phi/\partial z, \end{aligned}\right\} \tag{4.8}$$

where the dots denote derivatives with respect to time, and

$$\left.\begin{aligned} \dot{x} &= v_x, \\ \dot{y} &= v_y, \\ \dot{z} &= v_z. \end{aligned}\right\} \tag{4.9}$$

Now it is possible to write down an equation for f by using a standard property of differential equations by which the six first order ordinary differential equations (4.8), (4.9) are exactly equivalent to a single first order partial differential equation. This single equation is

$$\frac{\partial f}{\partial t} + v_x\frac{\partial f}{\partial x} + v_y\frac{\partial f}{\partial y} + v_z\frac{\partial f}{\partial z} + \frac{\partial \Phi}{\partial x}\frac{\partial f}{\partial v_x} + \frac{\partial \Phi}{\partial y}\frac{\partial f}{\partial v_y} + \frac{\partial \Phi}{\partial z}\frac{\partial f}{\partial v_z} = 0. \tag{4.10}$$

It is shown in textbooks on differential equations† that such an equation is precisely equivalent to the equations

$$\frac{dt}{1} = \frac{dx}{v_x} = \frac{dy}{v_y} = \frac{dz}{v_z} = \frac{dv_x}{\partial\Phi/\partial x} = \frac{dv_y}{\partial\Phi/\partial y} = \frac{dv_z}{\partial\Phi/\partial z}, \tag{4.11}$$

which are in turn exactly equivalent to the equations (4.8) and (4.9).

The equivalence of these sets of equations is in the sense that, whilst the equations (4.11) describe the path of a star in the six-dimensional space (x,y,z,v_x, v_y,v_z), the surfaces $f=$ constant in that space contain the paths of the stars. Thus, f *is constant following the motion of the stars*. This is then a special case of a familiar

† See for example I.N.Sneddon, *Elements of Partial Differential Equations*, McGraw-Hill.

theorem in statistical mechanics known as *Liouville's theorem*. This equivalence can be shown readily as follows. Compare the value of f at the point (x,y,z,v_x, v_y,v_z) at time t with its value at the nearby point $(x + \delta x, y + \delta y, z + \delta z, v_x + \delta v_x, v_y + \delta v_y, v_z + \delta v_z)$ occupied by the same star at $t + \delta t$. Then

$$f(x + \delta x, y + \delta y, z + \delta z, v_x + \delta v_x, v_y + \delta v_y, v_z + \delta v_z, t + \delta t)$$

$$-f(x, y, z, v_x, v_y, v_z, t)$$

$$= \frac{\partial f}{\partial t}\delta t + \frac{\partial f}{\partial x}\delta x + \frac{\partial f}{\partial y}\delta y + \frac{\partial f}{\partial z}\delta z + \frac{\partial f}{\partial v_x}\delta v_x + \frac{\partial f}{\partial v_y}\delta v_y + \frac{\partial f}{\partial v_z}\delta v_z$$

$$= \delta t\left[\frac{\partial f}{\partial t} + v_x\frac{\partial f}{\partial x} + v_y\frac{\partial f}{\partial y} + v_z\frac{\partial f}{\partial z} + \frac{\partial \Phi}{\partial x}\frac{\partial f}{\partial v_x} + \frac{\partial \Phi}{\partial y}\frac{\partial f}{\partial v_y} + \frac{\partial \Phi}{\partial z}\frac{\partial f}{\partial v_z}\right], \qquad (4.12)$$

using equations (4.11). The vanishing of the right hand side of (4.12) is then exactly equivalent to equation (4.10) as we have stated, showing that f is indeed constant following the motion of a star. If collisions are important, the value of f is changed by collisions between individual stars, whose speeds and directions of motion are changed. It is shown in textbooks on the kinetic theory of gases that an additional term $(\partial f/\partial t)_{\text{collisions}}$ must be added on the right hand side of equation (4.10), where the precise expression for $(\partial f/\partial t)_{\text{collisions}}$ depends on the forces acting between the particles. If a collisional system is a steady state, f must not be changed by collisions so that $(\partial f/\partial t)_{\text{collisions}}$ vanishes and equation (4.10) is still satisfied. This means that the Maxwellian distribution is one solution of (4.10).

Constants of the motion of a star

Having given what is in effect a plausibility argument for (4.10) being the correct equation for f, I may now pursue my question and ask what constraints the equation places on the form of f appropriate to a galaxy and in particular what time-independent fs are allowed. If I think of equations (4.11) as being six first order ordinary differential equations for x,y,z,v_x,v_y,v_z as functions of t, I can imagine those equations being solved in principle. Such solutions will involve six constants of integration, one for each equation, and these constants will be determined if the position and velocity of the star are specified at some time t.

Another way of saying this is that six functions of x,y,z,v_x,v_y,v_z,t must be constant on the trajectory of a star. It is then possible to demand that t be expressed in terms of x,y,z,v_x,v_y,v_z and one of the constants and that this expression be used to eliminate t from the other five functions. If this is done, there are five functions of x,y,z,v_x,v_y,v_z which are constant along the path of a star. These five time-independent constants of the motion can be written $I_1(x,y,z,v_x,v_y,v_z), \ldots, I_5(x,y,z,v_x,v_y,v_z)$ and in a time-independent stellar system only these five quantities or any combinations of them can be constant following the motion of a star. But we have already seen that f must be such a constant. This implies that

$$f = f(I_1, \ldots, I_5) \tag{4.13}$$

and gives us an idea of the possible forms that f can have in a stellar system provided that we can evaluate the integrals $I_1, \ldots, I_5$. Can we do this?

Constants of motion in spherical, axisymmetric and triaxial galaxies

I consider first a spherical system. This discussion is appropriate for a spherical (E0) galaxy and for some globular star clusters. There are four fairly obvious constants of the motion of a star in such a system if its overall properties are not changing with time, which implies specifically that Φ is time-independent. The first constant, which does not depend on the spherical symmetry of the system, is the total energy per unit mass of the star, kinetic plus gravitational potential. This gives us

$$I_1 \equiv \tfrac{1}{2}(v_x^2 + v_y^2 + v_z^2) - \Phi. \dagger \tag{4.14}$$

The constancy of I_1, which depends only on the vanishing of $\partial\Phi/\partial t$ and not on the symmetry of the system, is easily verified. Thus

$$\frac{dI_1}{dt} = \frac{\partial I_1}{\partial v_x}\frac{dv_x}{dt} + \frac{\partial I_1}{\partial v_y}\frac{dv_y}{dt} + \frac{\partial I_1}{\partial v_z}\frac{dv_z}{dt} + \frac{\partial I_1}{\partial x}\frac{dx}{dt} + \frac{\partial I_1}{\partial y}\frac{dy}{dt} + \frac{\partial I_1}{\partial z}\frac{dz}{dt}$$

$$= v_x\frac{\partial\Phi}{\partial x} + v_y\frac{\partial\Phi}{\partial y} + v_z\frac{\partial\Phi}{\partial z} - \frac{\partial\Phi}{\partial x}v_x - \frac{\partial\Phi}{\partial y}v_y - \frac{\partial\Phi}{\partial z}v_z = 0.$$

The three other obvious integrals arise because in a spherical system the total gravitational force on any star must act towards the centre of the system and does not therefore exert any torque about that point. This means that the three components of angular momentum per unit mass of the star about the centre of the system must be constant giving, in cartesian co-ordinates,

$$I_2 \equiv xv_y - yv_x, I_3 \equiv yv_z - zv_y, I_4 \equiv zv_x - xv_z. \tag{4.15}$$

Most galaxies are not spherical but many are instead axisymmetric, at least to a first approximation. As far as their mass distribution is concerned, most ordinary spiral galaxies only depart slightly from symmetry about an axis and this is also true for lenticular galaxies. Barred spirals clearly depart significantly from axisymmetry at least in their central regions. It used to be thought that all elliptical galaxies were axisymmetric oblate spheroids but it is now clear that many are triaxial. In an axisymmetric galaxy the energy of a star is still a constant of the motion, but now there is only one component of angular momentum which is conserved. The force acting on a star is in general no longer directed towards the

---

† Note that some authors, including Mihalas and Binney, use a definition of Φ with the opposite sign. My sign is chosen for consistency with the convention normally used in a study of stellar structure.

centre of the galaxy but it still necessarily intersects the axis of symmetry. This means that the angular momentum about that axis (the axis of rotation of the entire galaxy, if the flattening of the galaxy is caused by rotation) is constant. This gives an integral which can be written in cylindrical polar co-ordinates $(\tilde{\omega}, \phi, z)$

$$I_2 \equiv \tilde{\omega} v_\phi, \tag{4.16}$$

where this I_2 is the angular momentum per unit mass about the z axis just as in (4.15). If a galaxy is triaxial, it possesses no axis of symmetry and no component of angular momentum is constant. As a result the only obvious integral is the energy I_1.

Given the existence of these integrals I can now say that, in the case of a *spherical* galaxy, a *possible* form for f is

$$f = f(I_1, I_2, I_3, I_4). \tag{4.17}$$

There are, however, two comments which I must make on this. The first, which relates to the possible existence of a fifth integral I_5, will be deferred for a moment. The second is concerned with the requirement that the distribution function f has sufficient symmetry that a galaxy containing stars with that distribution is spherical. Thus, the requirement that the stars produce the gravitational field in which they themselves move is expressed by *Poisson's equation*, which in spherical polar co-ordinates in a spherical system can be written

$$\frac{1}{r^2}\frac{d}{dr}\left(r^2\frac{d\Phi}{dr}\right) = -4\pi G\rho = -4\pi G \int f dv_x dv_y dv_z. \tag{4.18}$$

The form (4.17) for f will only make sense if the resulting density, ρ, has spherical symmetry and it can be shown that this will be true provided f has the special form

$$f = f(I_1, I_2^2 + I_3^2 + I_4^2). \tag{4.19}$$

A particular form of f which satisfies this constraint is the Maxwellian distribution, which depends on I_1 alone, but there are many other possibilities. It is possible to build models of axisymmetric galaxies using fs of the form $f(I_1, I_2)$, but no triaxial galaxy can be constructed with $f(I_1)$. Such a distribution function, which is symmetrical in v_x, v_y, v_z, must lead to a spherically symmetrical mass distribution and hence to a spherical galaxy.

In writing down equation (4.18) I have assumed that the stars provide all the mass of the galaxy. However, as I shall explain in the next chapter, there is a large amount of hidden mass in galaxies and this hidden mass is probably in the form of weakly interacting elementary particles rather than stars. This means that there must be an additional term on the right hand side of (4.18) provided by the density of the hidden matter. This should not affect any argument about the form (4.19) for the distribution function. It is highly unlikely, if not impossible, that separate asymmetrical distributions of stars and weakly interacting particles could add together to give a spherical galaxy.

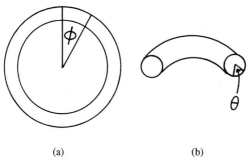

(a)   (b)

Figure 61.

Isolating and non-isolating integrals

I must now ask about the remaining integral of motion in the spherical case and the remaining three or four integrals in the axisymmetric and triaxial cases respectively. Can they be found and if so what form do they have? I will in fact concentrate on the non-spherical cases which are of more general interest. It can first be said that in the general case no additional integral has been found which has a simple analytical form. The integrals clearly must exist but at this point I should introduce a distinction between what are known as *isolating* and *non-isolating* integrals. Consider specifically the axisymmetric case. I_1 = constant defines a surface (five-dimensional) in the six dimensional space x,y,z,v_x,v_y,v_z. I_2 = constant is a similar surface and they intersect in a four-dimensional surface; let us call this surface S_4. Suppose a third integral of motion is I_3. Either the intersection of the surface I_3 = constant with S_4 is a three-dimensional surface S_3 or the surface I_3 = constant goes arbitrarily close to every point of S_4. In the former case I_3 (as well as I_1, and I_2) is called an *isolating integral*. In the latter case it is called a *non-isolating integral*. In this case it makes no sense to ask the distribution function to depend on I_3 because, once I_1 and I_2 are specified, f has the same value for all values of I_3. This leads us to what is known as *Jeans' theorem* in stellar dynamics; only isolating integrals must be used in f.

This may be illustrated by a specific example from a space of a smaller number of dimensions. Suppose the process of giving values to integrals has defined a surface which is a torus in three dimensions (fig. 61). As indicated in fig. 61 position on the torus can now be specified by two angles θ and ϕ. Suppose the next integral is

$$I = l\theta + m\phi = \text{constant}, \tag{4.20}$$

where l and m are known constants. If l/m is a rational number, the helix (4.20) closes after a number of circuits around the torus and I is an isolating integral; if l/m is irrational, the helix eventually goes arbitrarily close to every point on the torus and the integral is non-isolating.

I now return to the problem of axisymmetric galaxies. My previous questions about further integrals for axisymmetric systems can now be replaced by asking whether an arbitrary axisymmetric system possesses any further isolating integrals. As I have already said, no simple analytical integrals have been found in the general case. This means that they must be sought by directly integrating the equations of motion of a star with given values of energy and of angular momentum about the axis of symmetry and by looking for integral surfaces. This involves lengthy computations with large computers and no computations can be free from errors. If values of I_1 and I_2 are specified, the equations to be solved can be expressed in terms of $\tilde{\omega}$, $\dot{\tilde{\omega}}$ and z alone. When the equations are solved, it is possible to make a two-dimensional plot of the values of $\tilde{\omega}$ and $\dot{\tilde{\omega}}$ each time $z = 0$ and $\dot{z} > 0$, for example. If the resulting points form a closed curve in this two-dimensional space, which is known as a surface of section, this is evidence for the existence of a third isolating integral. If the points scatter throughout a finite region of the $(\tilde{\omega}, \dot{\tilde{\omega}})$ plane, this shows that there is no isolating integral. In fact, it is not uncommon for an isolating integral to exist for some pairs of values of I_1, I_2, but not for others. The computations indicate than an axisymmetric galaxy with a mass distribution similar to our Galaxy probably has a third isolating integral. I now explain why that is particularly interesting.

The third integral and the Galaxy

Whatever is the theoretical evidence for the existence of a third isolating integral of motion in flattened galaxies, there is good observational evidence either that such an integral does exist in our Galaxy or that one of the other assumptions which I have been making is not true. The reason for this is as follows. I have already described on page 28 a property of the random motions of the stars in the solar neighbourhood. This can be written as

$$<(v_\phi - v_{\phi 0})^2> \approx <v_z^2> \approx 0.4<v_{\tilde{\omega}}^2>. \tag{4.21}$$

Now suppose that f for our Galaxy is a function of I_1, I_2 alone so that

$$f = f(v_{\tilde{\omega}}^2 + v_\phi^2 + v_z^2 - 2\Phi, \tilde{\omega}v_\phi). \tag{4.22}$$

$v_{\tilde{\omega}}$ and v_z enter symmetrically in (4.22) so that there is no way in which (4.21) can be satisfied. If f had the form (4.22), we should have

$$<v_{\tilde{\omega}}^2> = <v_z^2>. \tag{4.23}$$

If there is a third isolating integral in which $v_{\tilde{\omega}}$ and v_z enter asymmetrically, (4.22) will no longer be true and, if the third integral has an appropriate form, it is possible to see how (4.21) can hold. We deduce either that there is a third integral or that departures from a steady state or from axial symmetry are important. Although in a real galaxy there must be some departures from a steady state and axial symmetry, the most plausible explanation for the large difference between (4.21) and (4.23) is that there is a third integral.

The theoretical study of the equilibrium of triaxial systems is much more difficult. After a value has been specified for I_1, equations in five variables remain. In addition it is likely that models of triaxial galaxies require two additional isolating integrals. Despite this, progress has been made in discovering such integrals and, thus, in demonstrating that it is possible to have time-independent triaxial systems, resembling both barred spirals and triaxial ellipticals. There are some particular forms of triaxial potential for which three analytical isolating integrals can be found. Although real galaxies may not be precisely similar to model galaxies which can be constructed using these potentials, they provide a proof that time-independent triaxial systems can exist.

One further comment can be made about the numerical study of integrals. A star has perhaps completed about fifty orbits in a galactic lifetime. If it is necessary to integrate for more than this number of orbits to determine whether or not an integral is isolating, is it relevant whether it is or not? A final point is that it is necessary for equilibrium distribution functions to be stable, if they are to be relevant. An unstable distribution could change significantly in the time required for a few stellar orbits in the galaxy or, perhaps more realistically, would never have been achieved. In recent years a considerable effort has gone into investigating whether or not distribution functions proposed for galaxies and star clusters are stable.

The distribution function in a dense stellar system

Having indicated that I do not expect the velocity distribution function of stars in a galaxy to be Maxwellian and having indicated briefly what other possibilities there are, I will not discuss the general problem of the equilibrium of stars in a galaxy any further. I will conclude this discussion by making a few remarks about the evolution of the velocity distribution in the dense central regions of a galaxy or of a dense star cluster. In such a case collisions cannot be regarded as unimportant and there is likely to be a tendency for the distribution to become approximately Maxwellian. There is still an important distinction between the situation inside a star where the mean free path between collisions for particles is a minute fraction of a stellar radius and in a star cluster where the mean free path will still usually be very much larger than the cluster radius, so that a star will oscillate across the cluster many times between successive collisions. In a star the temperature, or equivalently the average energy of the particles, is much higher near the centre than near the surface. In a star cluster the large mean free path prevents this concentration of energy near the centre and the cluster will approach what is effectively an isothermal state.

There is, however, no bounded equilibrium of an *isothermal gas sphere*. This can be seen as follows. In any self-gravitating system in equilibrium, the total kinetic energy and potential energy are approximately equal. Thus

$$GM^2/R \approx \tfrac{1}{2}M\langle v^2\rangle, \tag{4.24}$$

where M and R are the total mass and radius of the system and $\langle v^2\rangle$ is the average value of v^2 for the molecules (see Appendix 2). Equation (4.24) says that the typical molecule has about escape velocity. In a star the central temperature is very much higher than the surface temperature so that particles near the centre are moving much faster than those near the surface. As a result (4.24) can be satisfied without the system dispersing. In an isothermal system, the average particle speed is the same everywhere and many particles near the surface can escape. This implies that, as a dense star cluster relaxes towards equilibrium, stars will escape from it. At the same time the remaining system will become more compact and as a consequence it will evolve even faster. When it is realised that there is a wide variety of stellar mass and that collisions tend to equalise the distribution of kinetic energy, it is clear that low mass stars will acquire the highest velocities and that they will be most likely to escape. It does, indeed, appear that this process has already occurred in globular clusters, which seem to be deficient in low mass stars. It is very probable that they were present initially but that they have since escaped from the clusters and have become the individual high velocity stars which we observe to be associated with the halo of the Galaxy. Another interesting process which can occur in dense stellar systems is the formation through collisions of tight binary stars with a release of gravitational binding energy which can influence the evolution of the whole system. I shall discuss this process of dynamical evolution of stellar systems further in Chapter 8 when I shall also see that corresponding effects can be important in the evolution of clusters of galaxies.

The motion of individual stars in the Galaxy

I will now conclude this chapter by discussing what I have previously called a test particle problem. We have seen in Chapter 2 that the main motion of stars in the Galaxy is one of rotation about the galactic centre. Individual stars have small random velocities in addition to their rotational velocities and it is of interest to ask how far apart two stars which are at present close together can get in their future motion or equivalently what fraction of the volume of the Galaxy can be explored by an individual star. I shall first consider motion in the galactic plane and subsequently motion perpendicular to the galactic plane.

Consider two stars which initially coincide in position at the Local Standard of Rest. Suppose that one star is moving with pure circular motion with the velocity of the LSR but that the other star as well as sharing the circular velocity has an additional small velocity in the $\tilde{\omega}$ direction (fig. 62). I suppose that this $\tilde{\omega}$ velocity is sufficiently small that the region explored by the star is such that I can take the rotation curve as linear and described by Oort's constants A and B.

The motion of the star can be obtained by writing down the equation of motion in the $\tilde{\omega}$ direction together with the law of conservation of angular momentum about the axis of rotation of the Galaxy. When the star which is initially at radius R_0 is at a point where the value of $\tilde{\omega}$ is

$$\tilde{\omega} = R_0 + \xi, \tag{4.25}$$

the radial equation of the equation of motion has the form

$$\ddot{\xi} - (R_0 + \xi)\dot{\phi}^2 = \partial\Phi/\partial\tilde{\omega}, \tag{4.26}$$

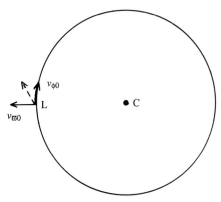

Figure 62.   The velocity components of a star relative to the Local Standard of Rest.

where I have noted that the radial component of acceleration in cylindrical polar coordinates is

$$a_{\tilde{\omega}} = \ddot{\tilde{\omega}} - \tilde{\omega}\dot{\phi}^2. \tag{4.27}$$

The condition of conservation of angular momentum about the axis of rotation can further be written

$$(R_0 + \xi)^2\dot{\phi} = R_0^2\omega_0 = R_0v_{\phi 0}, \tag{4.28}$$

where ω_0 and $v_{\phi 0}$ are the angular velocity and circular velocity of the LSR. If I use equation (4.28) to eliminate $\dot{\phi}$ from equation (4.26) I obtain

$$\ddot{\xi} - [R_0^2v_{\phi 0}^2/(R_0 + \xi)^3] = \partial\Phi/\partial\tilde{\omega}. \tag{4.29}$$

Now $\partial\Phi/\partial\tilde{\omega}$ at the radius $R_0 + \xi$ is related to the circular velocity at that radius and this in turn is related to $v_{\phi 0}$ and to Oort's constants A and B. Thus

$$v_{\text{circ}}^2(R_0 + \xi)/(R_0 + \xi) = -\partial\Phi/\partial\tilde{\omega} \tag{4.30}$$

and

$$v_{\text{circ}}(R_0 + \xi) = v_{\phi 0} - (A + B)\xi, \tag{4.31}$$

the latter following from equations (2.24) and (2.25) defining A and B and (4.25). If equations (4.29), (4.30) and (4.31) are combined, I have

$$\ddot{\xi} - [R_0^2v_{\phi 0}^2/(R_0 + \xi)^3] = - [v_{\phi 0} - (A + B)\xi]^2/(R_0 + \xi). \tag{4.32}$$

I can now obtain an equation governing small displacements from radius R_0 by expanding the second and third terms in equation (4.32) and by retaining only linear terms in ξ. If $v_{\phi 0}$ is written in terms of A and B by use of (2.24) and (2.25), there finally results

$$\ddot{\xi} + [-4B(A - B)]\xi = 0, \tag{4.33}$$

or

$$\ddot{\xi} + \varkappa^2 \xi = 0, \tag{4.34}$$

where, because B is negative and $(A - B)$ is positive, $\varkappa^2$ is positive. Equation (4.34) is an equation for simple harmonic motion, showing that the star oscillates about the radial position of the LSR with a period

$$P_{\bar{\omega}} = 2\pi/\varkappa = \pi/[-B(A - B)]^{1/2}. \tag{4.35}$$

If I suppose that at $t = 0$, $v_{\bar{\omega}} = v_{\bar{\omega}0}$, I can write the solution of (4.34) as

$$\xi = (v_{\bar{\omega}0}/\varkappa) \sin \varkappa t \equiv \xi_0 \sin \varkappa t. \tag{4.36}$$

Epicyclic motion

I can now use the conservation of angular momentum to determine the separation of the two stars in the ϕ direction. At any time the angular velocity of the star from equations (4.28) and (4.36) has the approximate form

$$\dot{\phi} = R_0 v_{\phi 0}/(R_0 + \xi)^2$$
$$\approx \omega_0 - (2v_{\phi 0}v_{\bar{\omega}0}/\varkappa R_0^2) \sin \varkappa t. \tag{4.37}$$

We can see that the star tends to lag behind or gain on the angular motion of the LSR, the star lagging when it is further from the galactic centre than the LSR. The difference in angular velocity is

$$\Delta\dot{\phi} = -(2v_{\phi 0}v_{\bar{\omega}0}/\varkappa R_0^2) \sin \varkappa t. \tag{4.38}$$

The difference in tangential velocity is then approximately $R_0\Delta\dot{\phi}$ and the expression for this can be integrated to give the tangential separation (η) of the star and the LSR as

$$\eta = [2v_{\phi 0}v_{\bar{\omega}0}/\varkappa^2 R_0][\cos \varkappa t - 1]$$
$$\equiv \eta_0[\cos \varkappa t - 1]. \tag{4.39}$$

It is now possible to combine equations (4.36) and (4.39) to obtain the equation

$$(\xi^2/\xi_0^2) + [(\eta + \eta_0)^2/\eta_0^2] = 1, \tag{4.40}$$

which is the equation of an ellipse. This indicates that a star moves in an elliptic orbit relative to the Local Standard of Rest or alternatively in an epicyclic orbit in a non-rotating frame. This is illustrated in figs. 63 and 64.

Although the precise expressions which I have obtained here in terms of local values of Oort's constants can only be strictly valid for small departures from the circular motion, they give a good first idea of more general motion. It is thus possible to estimate how far towards and away from the galactic centre the Sun and other stars in the solar neighbourhood have moved in the past and will move in the future. I defer any discussion of the numerical values until I have discussed motion perpendicular to the plane.

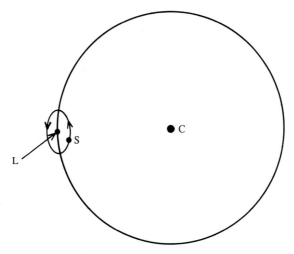

Figure 63.   The elliptical motion of a star S around the local standard of rest.

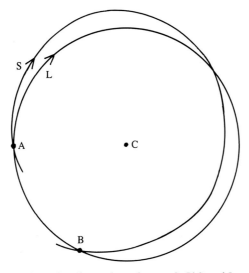

Figure 64.   The epicyclic motion of a star S. If S and L coincide at A, they next coincide at B.

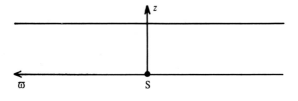

Figure 65.   A section of the galactic disk near the Sun, S.

Motion perpendicular to the galactic plane

The general motion of a star includes a component perpendicular to the galactic plane. For large amplitude deviations from circular motion, the motion in the plane and that perpendicular to the plane cannot be considered separately, but in the approximation that I am making in which I keep only those terms which are linear in the displacement from equilibrium the two components of motion do decouple. To discuss the motion of stars in the solar neighbourhood in the direction perpendicular to the galactic plane, I shall note that, as I shall discuss further in Chapter 6, because the Galaxy is very thin compared to its radius its properties vary very much more rapidly in the z direction than in the radial direction. This means that to a first approximation I can regard the disk in the solar neighbourhood as infinite and plane parallel with properties varying only in the z direction (fig. 65). I consider a co-ordinate system rotating with the LSR and study the vertical motion of the stars. I thus suppose that the stars move up and down through the disk under the influence of a gravitational field g_z. To determine the extent and period of these motions I shall need to have a value for g_z. How is it possible to determine this? In principle I do it by observing the stars oscillating up and down through the plane of the Galaxy. However, such an observation would take much more than 10^7 years so that it is clearly not practicable. Instead, if I suppose that the Galaxy is in a steady state, I can use the density and velocities of the stars as a function of height above and below the galactic midplane to obtain g_z.

The gravitational field perpendicular to the galactic disk

Let me first consider the principle of the method. We are situated almost exactly in the midplane where the vertical component (g_z) of the gravitational field vanishes. We can observe stars in our neighbourhood and in particular observe their distribution of velocities perpendicular to the galactic plane (henceforth called vertical velocities). Any individual star moving vertically upwards comes under the influence of a gravitational field downwards and eventually its direction of motion is reversed and it returns to the midplane. The greater is its vertical velocity when it is near to the Sun, the further it will travel. If I knew the distribution of g_z with height I should be able to calculate how far each star would travel. Instead what I can observe is the number density of stars as a function of height z above the galactic plane. It is then possible to show that, assuming that the stellar distribution is in a steady state, I can deduce the run of gravitational field with height from the distribution of stellar numbers.

If I assume that the vertical motion of stars is decoupled from their horizontal motion, stars moving in the vertical direction have an energy integral of the form

$$\tfrac{1}{2}v_z^2 - \Phi(z) = \tfrac{1}{2}v_{z0}^2 - \Phi(0), \tag{4.41}$$

where v_z is the z velocity of the star, v_{z0} its velocity at $z = 0$ and $\Phi(z)$ the gravitational potential, with

$$g_z = d\Phi/dz. \tag{4.42}$$

In a realistic galactic potential (4.41) will be true to a close approximation provided $z \ll \tilde{\omega}$. I can observe the distribution of velocities of a group of stars in the solar neighbourhood. Let me first suppose that I find that the distribution has the particularly simple form

$$f(v_{z0}) \propto \exp(-l^2 v_{z0}^2), \tag{4.43}$$

where l is a constant. I then know from the general properties of equilibrium distribution functions described earlier in this chapter that f will be a function of the energy integral (4.41), so that at any height z I must have

$$f(v_z) \propto \exp(-l^2 v_z^2 + 2l^2\Phi(z) - 2l^2\Phi(0)), \tag{4.44}$$

with the same constant of proportionality in (4.43) and (4.44). Here I am simply using the property that in equilibrium f must be constant following the motion of the stars which I established on page 95. I can now integrate (4.43) and (4.44) over velocities $(-\infty < v_z < \infty)$ to obtain the total stellar densities at heights 0 and z and can see that

$$n(z)/n(0) = \exp 2l^2(\Phi(z) - \Phi(0)), \tag{4.45}$$

where $n(z)$ is the number of stars per unit volume at height z. Then if $n(z), n(0)$ and l can be observed, I can obtain a value for $\Phi(z) - \Phi(0)$; as usual only this difference in potential has any significance. If this can be repeated for a succession of values of z, I can then use (4.42) to give values for g_z.

The observed distribution of stellar velocities near the Sun is not quite as simple as (4.43), although it has rather similar properties, otherwise I would not have used it as an illustrative example. Obviously the procedure which has been described can be followed for a different functional form although the relation between $n(z)$ and Φ will not be as simple as (4.45) and the derivation of g_z will be more complicated. A better representation of the observed distribution function than a single exponential is a sum of exponentials with different values of l. The different values of l, for example, correspond to stars of different spectral types which occupy disks of different thicknesses, as has been explained in Chapter 2. Use of such an expression has given the results for g_z shown in fig. 66. It can be seen that close to the galactic midplane the value of $|g_z|$ rises almost linearly with z but that the rate of increase reduces later. There are important difficulties in this procedure which requires an accurate knowledge of the distances of stars above the midplane and of their number density. For this reason different authors obtain a variety of values for the slope of the relation between $|g_z|$ and z.

In Chapter 6 I shall relate the variation of gravitational potential to the density of gravitating matter through Poisson's equation in the simplified form

$$d^2\Phi/dz^2 = -4\pi G\rho, \tag{4.46}$$

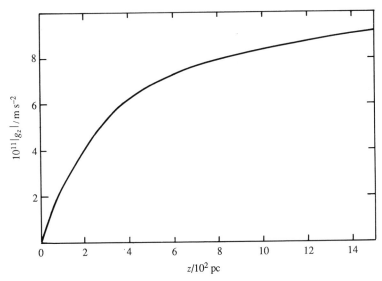

Figure 66. The magnitude of the z component of the gravitational field as a function of distance from the galactic midplane.

where ρ is the total matter density, but in this chapter I shall simply use the observed value of g_z to discuss the motions of stars perpendicular to the galactic disk. For distances less than 100 pc I can write

$$g_z = -\lambda z \tag{4.47}$$

where $\lambda \approx 10^{-29}$ s^{-2}. This implies that stars making small excursions from the galactic midplane move according to the equation

$$\ddot{z} + \lambda z = 0, \tag{4.48}$$

that is, with simple harmonic motion. With the value of λ given above, the time taken for a star to execute one oscillation is about 6×10^7 years and the distance above the midplane travelled for an initial velocity of 10 km s^{-1} is about 100 pc, with the distance being proportional to the velocity, at least for smaller initial speeds.

If the star's velocity in the midplane is high enough to take it outside the thin disk, it moves into a region in which the gravitational field is less strong than the linear value and both the period of oscillation and the amplitude of motion are greater than given by the simple formula. Of course, the oscillation is no longer simple harmonic and eventually the assumption that the z motion is decoupled from the motion in the galactic plane becomes completely invalid. This is particularly true for halo objects such as the globular clusters and high velocity stars which are at present in the solar neighbourhood. The orbital period for such halo objects is a few times 10^8 years.

Amplitude of the solar motion

I now return to the discussion of stellar motions in the galactic plane. As I have explained in Chapter 2, there continue to be significant uncertainties in the values of A, B, $v_{\phi 0}$ and R_0. If I take $v_{\phi 0} = 220$ km s^{-1} and $R_0 = 8.5$ kpc, the period of galactic rotation is $\approx 2.4 \times 10^8$ yr. Twenty years ago the best estimates were $v_{\phi 0} = 250$ km s^{-1} and $R_0 = 10$ kpc giving a slightly higher period of 2.5×10^8 yr. The values of $v_{\phi 0}$ and R_0 imply a value for $A - B$ which is ≈ 26 km s^{-1} kpc^{-1}. To estimate $\varkappa$ and hence $P_{\tilde{\omega}}$, we need a value for B, which is the most uncertain of the parameters. $|B|$ is probably greater than 11 km s^{-1} kpc^{-1}. If I use this value,

$$P_{\tilde{\omega}} \approx 1.8 \times 10^8 \text{ yr}, \tag{4.49}$$

which is somewhat less than the value of 2×10^8 yr accepted twenty years ago. The epicyclic period, $P_{\tilde{\omega}}$, is also somewhat less than the rotation period. Near the Sun the period of vertical motion is about a third of the period of epicyclic motion. A star with a random radial velocity of about 10 km s^{-1} has an amplitude of motion of about 300 pc. It is now easy to combine the information about the motion perpendicular to the plane to see what will be the shape of the complete orbit of a star and to see how much of the volume of the Galaxy it will sample. Using the best estimates of the Solar motion relative to the Local Standard of Rest which have been given on page 28, it is possible to see that in its orbit the Sun travels a *total* radial distance of about 800 pc and a *total* vertical distance of about 160 pc (± 400 pc from the mean radial position which is outside R_0 and ± 80 pc from the galactic midplane).

Resonant orbits

In the discussion of the motion of the Sun I have assumed that the Galaxy is strictly axisymmetric. In fact, in our Galaxy and other spiral galaxies the spiral structure, which I have already discussed briefly on page 68, represents a small departure from axial symmetry. Each time a star passes through a spiral wave its motion is perturbed slightly. If the rotation period of the star, the period of epicyclic motion and the rotation period of the spiral pattern are not related in a simple manner, the disturbances in successive orbits are out of phase and the net effect on the motion of the star is unimportant. If, in contrast, the spiral pattern at a given radius $\tilde{\omega}$ varies with ϕ and t in the manner $\sin(m\phi + \omega_s t)$, and ω_s, the rotation frequency ω, and the epicyclic frequency $\varkappa$ are related by

$$m(\omega - \omega_s) = 0 \text{ or } \pm \varkappa, \tag{4.50}$$

the force is in phase and the amplitude of motion is increased. This is referred to as a *resonance*. These resonances occur at particular values of $\tilde{\omega}$. When $\omega = \omega_s$ we have a corotational resonance and the other cases are called Lindblad resonances. Similar resonances occur as a result of the distortion from axisymmetry in a barred spiral. Note that even if $\omega_s = 0$, there are resonances when $\omega = \pm \varkappa/m$. A discussion of the significance of the resonances is outside the scope of this book but they are believed to play a rôle in setting up a long-lived spiral pattern in the first place.

Summary of Chapter 4

In this chapter I have considered a galaxy to be composed of stars alone. Any star moves under the influence of the gravitational interaction of all the other stars and possibly of any hidden matter which may be present. Because their sizes are very much smaller than their separations, they can be regarded as point masses and they can be treated as a gas of stars using the methods of the kinetic theory of gases. Except in the central regions of galaxies and in dense star clusters, the mean time between collisions for an individual star is very much greater than the present ages of galaxies. This means that the gas of stars can be considered to be a gas in which no collisions take place. Because of the absence of collisions, there is no reason why the distribution function of stellar velocities should have the Maxwellian form which is usually found for laboratory gases in which collisions are frequent. Instead it is shown that the velocity distribution function in a galaxy must be a function of quantities which are constant following the motion of an individual star. For a galaxy in a steady state, these constants of the motion include energy and three components of angular momentum in a spherical system and energy and the angular momentum about the axis of symmetry in an axisymmetric system, whereas the energy is the only obvious constant in a triaxial system. If realistic models of non-spherical galaxies are to be constructed, additional constants must be found by numerical integration of the equations of motion.

If a stellar system is sufficiently dense, collisions between stars are important. Such collisions tend to equalise the distribution of kinetic energy both between stars of different masses and at different points in the system. The effect of this is to cause an escape, especially of the low mass stars, from the surface of the system and a contraction of the whole system, leading to a further increase in the importance of collisions.

Although the motion of all the stars in a galaxy can only be studied statistically, it is possible to discuss the motion of the star through the background of the other stars. The chapter ends with the discussion of the motion of a star in the solar neighbourhood whose velocity departs slightly from that of galactic rotation. It is shown that it oscillates in a simple harmonic fashion both about its mean radial distance from the galactic centre and about the galactic midplane. An estimate is made of the extent of the Sun's motion through the Galaxy.

5

Masses of galaxies

Introduction

Knowledge of the masses of galaxies is very important for several different reasons. In the first place we should like to know whether the visible matter in galaxies (where in the present context I use the word *visible* for all matter that has been observed by any method) makes the major contribution to the total mass of the galaxies or whether there is a significant amount of matter in forms which have not yet been detected. The second most significant reason is concerned with the overall density in the Universe; this determines, for example, whether a presently expanding Universe will subsequently re-contract. I shall discuss this further in Chapter 8.

The masses of double galaxies

The mass of any astronomical body can only be determined by observations of motion, either of the body as a whole or of internal motions, combined with the assumption that the motion is controlled by the force of gravity. In the simplest case the masses of the Sun and the planets and of double stars can be obtained by applying *Kepler's laws*. If two stars of masses M_1 and M_2 are moving in elliptic orbits of semi-major axes a_1 and a_2 about their common mass centre, it can be shown that (fig. 67)

$$M_1 a_1 = M_2 a_2 \tag{5.1}$$

and

$$P^2 = 4\pi^2 (a_1 + a_2)^3 / G(M_1 + M_2), \tag{5.2}$$

where P is the period of revolution. Clearly both masses can be obtained from (5.1) and (5.2) if a_1, a_2 and P can be observed. In practice what we observe even in the case of a binary star system is not a_1 and a_2 but the projection of a_1 and a_2 on the plane perpendicular to the line of sight from the observer to the double star

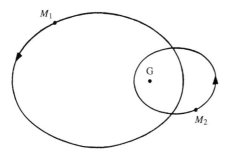

Figure 67. The elliptical orbits of two stars of masses M_1 and M_2 about their centre of mass G.

system, assuming in addition that the distance to the system is known, so that the apparent values of a_1 and a_2 can be converted into absolute values. This means that there is an uncertainty in the masses obtained. It is possible to apply a similar method to physically connected pairs of galaxies but here there are additional complications. The first is that galaxies are not such compact objects as stars and the separation of two galaxies forming a binary system (for example) may not be many times the radius of either. Because of this, the approximation of treating the two galaxies as point masses, which I do in a simple application of Kepler's laws, is somewhat dubious. This is nothing like as serious as the problem that we cannot observe the entire orbit of one galaxy about the other because the period of the motion is typically many hundreds of millions of years. All that we can observe is the apparent separation of the two galaxies and their components of velocity towards or away from us. If we believe that we have a good estimate of the distance to the galaxies we can convert the apparent separation to a distance in kpc but this is still not the actual distance between the galaxies because we do not know the inclination of the line joining the two galaxies to the line of sight (see fig. 68). We also cannot immediately decide what part of the observed radial velocity of either galaxy represents motion of the whole system towards or away from us and what part is due to the relative velocities of the two galaxies.

If we choose galaxies which are well separated, the approximation of treating them as point masses may not be too seriously inaccurate and we can then obtain some information about their masses. The best results will be obtained either if two very similar masses form a binary system, so that their masses may be assumed to be approximately equal, or when one galaxy is obviously very much less massive than the other. In the former case, if the observed radial velocities are v_1 and v_2, the radial velocity of the centre of mass will be $(v_1 + v_2)/2$. In the latter case if v_1 is the velocity of the more massive galaxy it can also be taken to be the velocity of the centre of mass. Even in these cases we still do not have enough information to obtain exact values for the masses because we do not know how the orbits of the galaxies are projected on to the plane perpendicular to the line of sight. All we can really do is to obtain statistical information from many pairs of galaxies by assuming that the planes of their actual orbits are orientated in a random fashion.

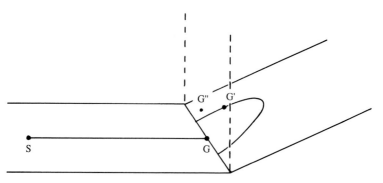

Figure 68. Relation of apparent orbit to true orbit. When a galaxy G' moves in the elliptical orbit shown about galaxy G, an observer at S sees G' projected to G".

Use of the Virial Theorem to obtain masses of single galaxies and groups of galaxies

I now turn to a method of estimating the masses of individual galaxies, which is also a method which can be used to estimate the total mass of clusters of galaxies. I suppose that I have a large number of particles which are interacting through the force of gravitation. The particles could be stars if I am considering a single galaxy or they could be galaxies if I am considering a cluster of galaxies. I then suppose that the system of particles is in a statistically steady state, of the type which I have described in the previous chapter. By this I mean that any changes in the overall properties of the system are occurring slowly so that there is essentially no change in these properties in the time taken for a single object to cross the system. If this assumption is made, it can be shown that the sum of the (negative) gravitational potential energy of the system and twice the kinetic energy must vanish. This is known as the *Virial Theorem* and it is proved in Appendix 2. It may be written

$$2T + \Omega = 0, \tag{5.3}$$

where T is the kinetic energy and Ω the potential energy. Now

$$T = \frac{1}{2} \sum_i mv_i^2, \tag{5.4}$$

where the summation is over all of the particles (stars or galaxies) and v_i is the speed of the particle of mass m_i. Alternatively I can write

$$T = \tfrac{1}{2}M\langle v^2 \rangle, \tag{5.5}$$

where M is the total of mass of the system and $\langle v^2 \rangle$ is an (appropriate) mean value of v_i^2. The gravitational potential energy is somewhat more complicated being given by

$$\Omega = -\sum_i \sum_{j \neq i} Gm_i m_j / r_{ij}, \tag{5.6}$$

where r_{ij} is the distance between masses m_i and m_j. The value of Ω for a complete system is dependent on the distribution of mass in the system but for a spherical (or nearly spherical) system it can certainly be written

$$\Omega = -\alpha GM^2/R, \tag{5.7}$$

where R is the radius of the system and α is a quantity whose value is of order unity but whose precise value depends on the spatial distribution of the mass in the system. For a system of uniform density $\alpha = 3/5$ and it increases above this value as the central density increases relative to the mean density. In a galaxy or a cluster of galaxies the central density will be much higher than the mean density, but, even for a central density about a thousand times the mean density, $\alpha \approx 3$.

Combining (5.5) and (5.7), I obtain

$$M = R\langle v^2\rangle/\alpha G, \tag{5.8}$$

as a formula for estimating the total mass of the system. Bearing in mind that $\alpha \approx 1$, (5.8) can be interpreted as saying that the average speed of particles in a self-gravitating system in equilibrium must be approximately the same as the escape velocity from the system; for a spherical mass of radius R, the escape velocity, v_{esc} is given by

$$v_{esc}^2 = 2GM/R. \tag{5.9}$$

We can see that v_{esc}^2 and $\langle v^2 \rangle$ are precisely the same for $\alpha = 2$. Of course the particles typically have higher than escape velocities when they are near the centre of the system and they are moving below escape velocity when they are near to the surface. If the average speed were much higher than that given by equation (5.8), the system would evaporate; if it were lower it would collapse. Equation (5.8) can be used to obtain the masses of spherical galaxies and of clusters of galaxies. In using it, all three of R, $\langle v^2 \rangle$ and α are only known very approximately, so that the masses obtained are themselves very uncertain. Once again only the radial velocities of the stars or galaxies can be observed and $\langle v^2 \rangle$ must be obtained by assuming that the mean square value of all three components of velocity is the same. The assumption is not unreasonable for a spherical system. Although the use of the Virial Theorem does give uncertain results, it is essentially the only method which can be used to obtain the masses of single elliptical (spherical) galaxies.

The determination of the masses of highly flattened galaxies

There is another method which can be used to determine the masses of highly flattened galaxies like our own. As I have already stated, the size of a spherical galaxy is determined by a balance between the random motions of the stars and the attractive force of gravity and that is what is really being expressed by equations (5.3) and (5.8). In the case of flattened galaxies the Virial Theorem (5.8) is still true but the kinetic energy of the stars now exists in two forms. There is the random kinetic energy arising because the stars at any given position in the

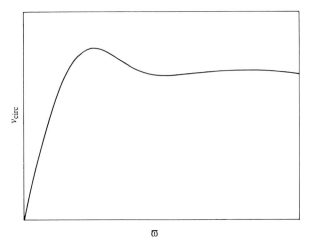

Figure 69.   A schematic galactic rotation curve.

galaxy are moving in different directions with different speeds. There is in addition the ordered motion of galactic rotation. As we saw in Chapter 2, in the solar neighbourhood in our Galaxy the ordered motion of galactic rotation is about 220 km s^{-1}, whereas the random speeds of most of the stars are no more than about a tenth of that. This means that the random kinetic energy is only of order one per cent of the ordered kinetic energy. It is this fact which causes the Galaxy to be as flat as it is; in the direction parallel to the axis of rotation only the random motion of the stars can balance the force of gravity, whereas the circular motion is much more significant than the random motion in the plane perpendicular to the axis of rotation.

Galactic rotation curves and galactic mass distributions

It is now possible to approach the problem of the mass distribution in, and the total mass of, highly flattened spiral galaxies by assuming that only the ordered motion of galactic rotation is important. Thus I consider an idealised galaxy in which all of the constituents, stars and gas, move in purely circular orbits about the centre of the galaxy. I have already used this approximation in Chapter 2 in discussing how our Galaxy rotates and in obtaining its *rotation curve*, the plot of circular velocity against distance from the galactic centre. In Chapter 3 I have discussed how a different technique can be used to obtain the rotation curves of suitably orientated nearby spiral galaxies. In either case, I can now use the rotation curves to obtain estimates of galactic masses. Although the two methods of obtaining rotation curves are very different the results obtained are qualitatively similar and the method of deducing masses is exactly the same however a rotation curve (fig. 69) has been obtained.

As the disk of such a galaxy is very thin, I can consider all of the stars to be moving in the galactic midplane. Then assuming that the galaxy is symmetrical

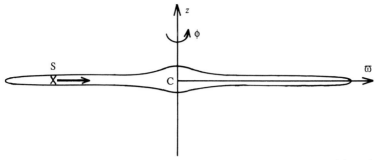

Figure 70.   A star, S, in the galactic midplane is acted on by a gravitational force in the direction of the arrow.

about its axis of rotation, or more specifically that spiral structure represents only a very small departure from symmetry in the overall mass distribution, we know that the force acting on any such star is directed towards the galactic centre and that (fig. 70)

$$v_\phi^2/\tilde\omega = -g_{\tilde\omega} \qquad (5.10)$$

where $g_{\tilde\omega}$, which is negative, is the gravitational field in the $\tilde\omega$ direction. Now the gravitational field is determined by the mass distribution and if we knew the gravitational field of the galaxy everywhere we might expect to be able to determine the mass distribution. If the matter density is ρ, the gravitational potential Φ satisfies *Poisson's equation*

$$\nabla^2\Phi \equiv \frac{1}{\tilde\omega}\frac{\partial}{\partial\tilde\omega}\left(\tilde\omega\,\frac{\partial\Phi}{\partial\tilde\omega}\right) + \frac{\partial^2\Phi}{\partial z^2} = -4\pi G\rho, \qquad (5.11)$$

where

$$g_{\tilde\omega} = \partial\Phi/\partial\tilde\omega, \quad g_z = \partial\Phi/\partial z. \qquad (5.12)$$

If I use equation (5.10) in the more general form

$$v_{circ}^2/\tilde\omega = -g_{\tilde\omega}, \qquad (5.13)$$

where v_{circ} is the velocity which appears in the galactic rotation curve, it is clear that I can use equation (5.13) to obtain $g_{\tilde\omega}$, or at least a value of $g_{\tilde\omega}$ averaged across the thickness of the galactic disk, from the rotation curve. This does not, however, give me enough information to use (5.11) to obtain ρ. To do this I need not only $g_{\tilde\omega}$ but also g_z throughout the galaxy. Although, as we have seen in Chapter 4, we can obtain a value of g_z in the solar neighbourhood in our Galaxy, we cannot obtain detailed information about it throughout our Galaxy, let alone in other galaxies. Nor can we assume that g_z is unimportant. As I shall discuss in Chapter 6, because the Galaxy is very thin, spatial gradients in the z direction are greater than in the $\tilde\omega$ direction and, in fact, it is the term $\partial^2\Phi/\partial z^2$ which determines ρ near to us, if we use equation (5.11). It is therefore clear that the rotation curve does not give us enough information for a galactic density distribution to be deduced unambiguously. However, some reasonably accurate information can be obtained from an indirect method which I now describe.

Simple models of galactic mass distribution

The method involves constructing simple models of a galactic mass distribution, the choice of model being based on the observed appearance of the galaxy. The model will contain some free parameters which can then be adjusted until the observed rotation curve is predicted to some acceptable approximation. A simple example of the approach can be given by reference to fig. 70. To a first approximation the galaxy in fig. 70 looks like a combination of a spherical system of small radius and a very flat spheroid of much larger radius. I can then ask whether the superposition of suitable spherical and spheroidal distributions of mass of the observed size will produce a rotation curve in agreement with that observed for the galaxy.

I shall discuss this method for our Galaxy almost immediately, but before I do so I should make explicit one important limitation to what we can learn about galactic masses from their rotation curves. A reliable curve can never be established right out to the edge of a galaxy. In our own Galaxy the rotation curve is not very well determined for distances from the galactic centre greater than that of the Sun but there are particular problems in determining the structure of a system from inside it. Even near the Sun, it is not certain whether v_{circ} is increasing or decreasing with $\bar{\omega}$.

In the case of other spiral galaxies, the flat rotation curves indicate that there must be mass at radii beyond those where reliable velocity measurements can be made. Let us suppose that the curve is known out to radius $\bar{\omega} = R$. The gravitational force acting on a star at distance R from the centre of a galaxy is mainly due to those stars and any invisible matter nearer to the centre of the galaxy than the star itself. This result would be exactly true for a spherical galaxy. It is also true if the galaxy is composed of concentric spheroidal shells of the same eccentricity and uniform density; these properties are discussed in Appendix 3 which is concerned with the gravitational fields due to spherical and ellipsoidal distributions of matter. Although a real galaxy will not possess these precise properties, in most cases the mass outside radius R in the midplane contributes very little to the force at radius R. Thus it may be possible to obtain reasonably reliable information about the mass of the galaxy within radius R but a study of the rotation curve out to radius R gives very little information about the mass contained within the galaxy at radii greater than R.[†] This means that the masses which are deduced must be regarded as lower limits to actual masses. This is important because there are very strong suggestions that galaxies including our own might have massive low density haloes, whose radii may be significantly greater than that of the easily visible galaxy. This will be discussed further below

† Or possibly more precisely the mass contained in spheroids whose semi-major axes are greater than
   R.

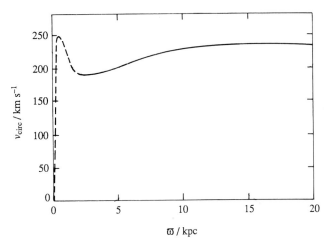

Figure 71.   The observed rotation curve for the Galaxy.

when I shall explain both why their existence has been suggested and how they might be detected other than through the rotation curve.

The mass of the Galaxy

I now turn to a discussion of the mass of our Galaxy as deduced from its rotation curve. I shall discuss a series of increasingly more realistic models of our Galaxy. Although I shall eventually discuss a model which will match the rotation curve to the accuracy with which it is itself determined, I first consider very crude models which can only reproduce a small fraction of the properties of the observed rotation curve (fig. 71). What we will subsequently see is that even the very crudest models give a reasonable estimate of the mass of the galactic disk within the solar radius R_0, even though they give a very poor fit to the overall galactic rotation curve and hence to the detailed mass distribution. This is somewhat reassuring as our main interest is in obtaining values for total masses of galaxies.

I consider first the crudest possible model. It is known that a significant fraction of the mass of the Galaxy is contained within the nucleus. I first assume that essentially all of the mass of the Galaxy is contained in a spherical nucleus of mass M_P; at the solar position this produces the same gravitational field as a point mass, hence the suffix P. I now ask what value M_P would need to have to produce the observed rotation velocity near the Sun. I have

$$GM_P/R_0^2 = v_{\phi 0}^2/R_0),\tag{5.14}$$

where I must use $R_0 = 8.5$ kpc and $v_{\phi 0} = 220$ km s^{-1}. When these values are inserted into (5.14), I obtain

$$M_P = 0.9 \times 10^{11} M_\odot.\tag{5.15}$$

Although this point mass gives the correct rotation velocity at the Sun, it gives a very poor fit to the complete rotation curve. In particular the maximum value of

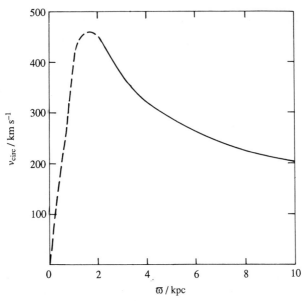

Figure 72.   The rotation curve produced by a point mass which gives the current estimate of the circular velocity for the Sun at a radius of 8.5 kpc.

the rotation velocity is far too high and near the Sun the velocity falls off much more rapidly with distance than is observed. At any radius outside the point mass we have

$$GM_p/\varpi^2 = v_{\mathrm{circ}}^2/\varpi, \qquad (5.16)$$

which gives the rotation curve shown in fig. 72. At the solar position, this model gives values of the Oort constants which are

$$A = 19.41 \text{ km s}^{-1} \text{ kpc}^{-1}, \quad B = -6.47 \text{ km s}^{-1} \text{ kpc}^{-1}, \qquad (5.17)$$

compared to the earlier generally accepted values $A = 15$, $B = -10$ in the same units. Despite the poor fit of the rotation curve, we shall see later that the value of the mass given by equation (5.15) probably differs by no more than 50 per cent from the actual mass contained within radius R_0 and this is the sort of error or uncertainty that has to be accepted in this subject at the present time. Note that any spherically symmetrical mass distribution which gives a circular velocity of 220 km s^{-1} at the Sun must have a mass of $0.9 \times 10^{11} M_\odot$ within radius R_0 although it can give a very different rotation curve from that of figure 72. I shall return to this point when I mention massive haloes again.

   The next simplest model of the Galaxy retains a point mass and recognises that there must be a significant amount of mass outside the nucleus. Observations of flattened spiral galaxies like our own indicate that the disk has a shape (as shown by its light distribution) which is approximately that of a very flat spheroid, apart from the approximately spherical nucleus. I shall therefore consider a model which has a point mass and a spheroidal mass and will initially assume that the

Figure 73.   A model of the Galaxy composed of a point mass (more realistically a small sphere) and a spheroid. The Sun is at **x**.

spheroid has uniform density. This is still unlikely to be a realistic model because the density of the spheroid should almost certainly fall off with distance from the centre of the Galaxy. It should, however, be a better model than the first one unless any significant fraction of the mass does not have associated light and has a more spherical distribution. Since a flattened distribution exerts a stronger gravitational field in the plane than a sphere of the same mass, the next estimate of the mass of the galaxy must be smaller than the first. Because any part of such a spheroid which extends beyond radius R_0 will not exert any gravitational force in the solar neighbourhood, I shall assume that the semi-major axis of the spheroid is R_0 (fig. 73). However, we must continue to bear in mind that a considerable fraction of the mass of the Galaxy may be situated outside this spheroid.

To discuss this model I need to know what is the gravitational field exerted at radius R_0 by such a spheroidal mass distribution. The derivation of the required expression is too complicated for the present book and I shall simply quote it.† Suppose that the equation of the elliptical cross section of the spheroid shown in fig. 73 is

$$(\tilde{\omega}^2/R_0^2) + [z^2/R_0^2(1 - e^2)] = 1, \tag{5.18}$$

where e is the eccentricity and $R_0(1 - e^2)^{1/2}$ the semi-minor axis of the ellipse; then the field towards the centre of the spheroid at $\tilde{\omega}$ is

$$-g_{\tilde{\omega}} = \frac{3GM_{\text{sph}}}{2R_0^3 e^3}(\sin^{-1} e - e(1 - e^2)^{1/2})\tilde{\omega}, \tag{5.19}$$

where M_{sph} is the total mass of the spheroid. Now the spiral galaxies are observed to be very flat so that to a first approximation I can put $e = 1$. If I do this, (5.19) becomes

$$-g_{\tilde{\omega}} = 3\pi GM_{\text{sph}}\tilde{\omega}/4R_0^3. \tag{5.20}$$

When I take account of the point mass as well, the rotation velocity of the Galaxy according to this model is given by

$$v_{\text{circ}}^2/\tilde{\omega} = GM_{\text{P}}/\tilde{\omega}^2 + 3\pi GM_{\text{sph}}\tilde{\omega}/4R_0^3. \tag{5.21}$$

It is now possible to fit the rotation curve given by equation (5.21) to some properties of the observed rotation curve. It is fairly easy to see that no choice of values of M_{P} and M_{sph} will give the full complexity of the observed curve. As there are just these two free parameters in (5.21) I can fit two properties of the observed rotation curve. The simplest properties to fit are the observed values of Oort's

† The gravitational fields due to spheroidal distributions of matter are discussed in W. D. MacMillan, *The Theory of the Potential*, Dover.

constants (A and B), and that is what I will do, although the best fit to the overall rotation curve would certainly be obtained by fitting observed properties at two distinct points, neither of them necessarily coinciding with the Sun. From equation (5.21) it is possible to obtain

$$v_{\phi 0}^2 = GM_P/R_0 + 3\pi GM_{sph}/4R_0 \tag{5.22}$$

and

$$v_{\phi 0}(dv_{circ}/d\bar{\omega})_{R_0} = -GM_P/2R_0^2 + 3\pi GM_{sph}/4R_0^2. \tag{5.23}$$

Now using the definitions (2.24) and (2.25) of Oort's constants A and B given on page 34 I have

$$(A - B)^2 = GM_P/R_0^3 + 3\pi GM_{sph}/4R_0^3 \tag{5.24}$$

and

$$-(A - B)(A + B) = -GM_P/2R_0^3 + 3\pi GM_{sph}/4R_0^3. \tag{5.25}$$

As we have seen in Chapter 3 the values of A and B are uncertain. If, for example, I take $A = 14$ km s^{-1} kpc^{-1} and $B = -11.9$ kms s^{-1} kpc^{-1} together with $R_0 = 8.5$ kpc and these are inserted into (5.24) and (5.25), values are obtained for M_P and M_{sph}. If this is done

$$M_P = 0.69 \times 10^{11} M_\odot \tag{5.26}$$

and

$$M_{sph} = 0.11 \times 10^{11} M_\odot, \tag{5.27}$$

giving a total mass within radius R_0 of

$$M = 0.80 \times 10^{11} M_\odot. \tag{5.28}$$

The mass obtained from this model differs only very slightly from the mass obtained using a point mass alone. The rotation curve is not very different as can be seen from fig. 74, where it has been assumed that the point mass has a radius of 2 kpc and that inside this radius it has uniform density. It is also clear that the rotation curve (5.22) still lacks many of the important properties of the observed rotation curve. In particular it does not have the two maxima and if, as I have arranged, it gives the correct circular velocity at the solar radius it predicts far too high a circular velocity much nearer to the galactic centre. Perhaps a better two parameter fit can be obtained by taking v_{circ} equal to 220 km s^{-1} at some smaller radius, say $R_0/4$, as well as at R_0. If that is done a significantly smaller mass estimate of $0.47 \times 10^{11} M_\odot$ is obtained. The model also gives a rotation velocity which is increasing rather rapidly at R_0, which certainly disagrees with observations.

Schmidt's model of the Galaxy

A further refinement of the above model, which was introduced by M. Schmidt, was to replace the spheroidal distribution above by one of non-uniform

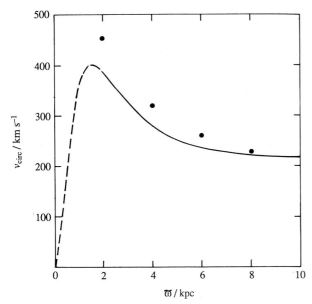

Figure 74. The rotation curve produced by a point mass and a uniform spheroid which gives satisfactory values for the Oort constants. The filled circles give the curve of fig. 72.

density. Although he also retained a point mass, this proved to be relatively unimportant in his model. This was possible because he obtained the high central mass concentration which the Galaxy certainly possesses, by allowing the density of the spheroidal distribution of mass to become very high near to the centre of the Galaxy. As well as allowing a variation of density inside the spheroid, Schmidt did not take $e = 1$ but determined e by a fit to observations. In fact, it will be seen later that his value of e scarcely differed from 1.

In order that the rotation curve at a given point should still be determined by the mass of the spheroid within that point, Schmidt supposed that the density was constant on spheroidal shells which were concentric with the Galaxy and which all had the same eccentricity (fig. 75). If the semi-major axis of any one of these spheroids is α, the density can then be expressed as $\rho(\alpha)$. Alternatively the mass within the shell can be written $M(\alpha)$. Once again I quote a result for the gravitational field. The gravitational field exerted by such a set of spheroidal shells at a point in the equatorial plane is

$$-g_{\tilde\omega} = 4\pi G\sqrt{(1-e^2)}\int_0^{\tilde\omega}\frac{\rho(\alpha)\alpha^2\,\mathrm{d}\alpha}{\tilde\omega\sqrt{(\tilde\omega^2-\alpha^2e^2)}} = G\int_0^{\tilde\omega}\frac{\mathrm{d}M(\alpha)}{\tilde\omega\sqrt{(\tilde\omega^2-\alpha^2e^2)}}. \quad (5.29)$$

It is now possible to consider an arbitrary function $\rho(\alpha)$. It proves convenient to consider $\rho(\alpha)$ to be expanded in powers of α and to use a series with a small number of terms to give a reasonable fit to the rotation curve. At first sight one would expect to use only positive powers of α because negative powers give an infinite density at $\alpha = 0$, at the galactic centre. However, if we are not concerned to fit the observed rotation curve very near to the galactic centre, where it is not in

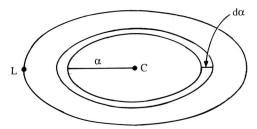

Figure 75.   Nested spheroids with the same eccentricity.

any case very well known, it is not inappropriate to use negative powers of α provided that they do not lead to an infinite total mass. As we shall see shortly, such a series is

$$\rho(\alpha) = (c_{-2}/\alpha^2) + (c_{-1}/\alpha) + c_0 + c_1\alpha + c_2\alpha^2 + \cdots, \tag{5.30}$$

where the cs are constants. I can now ask what will happen if any such term $c_n\alpha^n$ is inserted into the right hand side of (5.29) and its contribution to $g_{\tilde{\omega}}$ is calculated. It can first be seen that the integrand is not infinite at $\alpha = 0$ for $n \geqslant -2$, so that the series (5.30) is indeed suitable. For any value of n, the contribution to $g_{\tilde{\omega}}$ is

$$I_n \equiv 4\pi G \sqrt{(1 - e^2)} \int_0^{\tilde{\omega}} \frac{c_n\alpha^{n+2}\mathrm{d}\alpha}{\tilde{\omega}\sqrt{(\tilde{\omega}^2 - \alpha^2 e^2)}}. \tag{5.31}$$

To evaluate I_n, put $ae = \tilde{\omega}\sin\theta$ so that $\sqrt{(\tilde{\omega}^2 - \alpha^2 e^2)} = \tilde{\omega}\cos\theta$ and $\mathrm{d}\alpha = \tilde{\omega}\cos\theta\,\mathrm{d}\theta/e$. Then

$$I_n = [4\pi G \sqrt{(1 - e^2)}c_n\tilde{\omega}^{n+1}/e^{n+3}] \int_0^{\sin^{-1}e} \sin^{n+2}\theta\,\mathrm{d}\theta. \tag{5.32}$$

It is then a straightforward matter to evaluate the integral and it can be seen that, if I write

$$I_n = d_n\tilde{\omega}^{n+1}, \tag{5.33}$$

$$-g_{\tilde{\omega}} = (d_{-2}/\tilde{\omega}) + d_{-1} + d_0\tilde{\omega} + \cdots. \tag{5.34}$$

The value of the circular velocity can now be found from $g_{\tilde{\omega}}$ in the usual way and if I now also insert a point mass M_P I obtain

$$v_{\mathrm{circ}}^2 = (GM_P/\tilde{\omega}) + d_{-2} + d_{-1}\tilde{\omega} + d_0\tilde{\omega}^2 + \cdots, \tag{5.35}$$

where this formula is assumed to be valid outside the central mass concentration.

It is now possible to choose values of the constants $M_P, d_{-2}\cdots$, which will reproduce the observed rotation curve as well as possible. Schmidt found that, given the uncertainties of the observed curve, an adequate fit could be obtained with just three constants M_P, d_{-1} and d_1; this is just one more than in our previous model although the powers of $\tilde{\omega}$ involved are different. Because the galactic parameters are now believed to be different from those assumed by Schmidt I am modifying his fit by taking $R_0 = 8.5$ kpc and $v_{\phi 0} = 220$ km s^{-1} but by keeping the same shape for the rotation curve. From this fit M_P is determined directly and proves to be very small:

$$M_P = 0.049 \times 10^{11} M_\odot. \tag{5.36}$$

The values of d_{-1} and d_1 do not immediately give us values for c_{-1} and c_1 and hence of the contributions to the galactic mass due to them because the relation between the cs and ds involves the eccentricity e as is clear from equations (5.32) and (5.33). At this stage Schmidt used a result, which I shall obtain in the next chapter, which is that the density of matter in the solar neighbourhood is approximately $0.15 M_\odot$ pc^{-3}. This value of $\rho(a = R_0)$ enables an appropriate value of the eccentricity to be found which with our new values of R_0 and $v_{\phi 0}$ is 0.9986 (Schmidt's value was 0.9988). This is very close to unity as was to be expected and corresponds to a semi-minor axis of the spheroids which is five per cent of the semi-major axis. With this value of the eccentricity, the mass of the spheroidal distribution is

$$M_{sph} = 0.55 \times 10^{11} M_\odot, \tag{5.37}$$

giving a total mass interior to the Sun of

$$M = 0.60 \times 10^{11} M_\odot. \tag{5.38}$$

This value is once again less than the value given in (5.28) and each refinement of the galactic model has led to a reduction in the estimated mass. There is no reason to believe that a more detailed model will give an even smaller mass and it would be surprising if the mass interior to the Sun falls outside the range $0.5 \times 10^{11} M_\odot$. to $10^{11} M_\odot$ provided only that the currently accepted values of A, B and R_0 are essentially correct.

Schmidt also used what rudimentary information then existed about the rotation curve at radii outside the solar position to obtain an estimate of the mass outside the spheroid with semi-major axis R_0. The results obtained from this study must be regarded as less reliable than the results obtained above, but it is interesting that Schmidt obtained a value of the exterior mass which is almost exactly the same as his value of the mass corresponding to (5.38). Although the density of matter falls off very rapidly with distance there is an extremely large volume containing this matter. Schmidt thus estimated that the total mass of the Galaxy is about twice the mass inside the solar radius. It seems clear that the total mass of the galaxy is at least of order $1.5 \times 10^{11} M_\odot$ as I have stated earlier in the book but it could be much larger if it possesses a massive halo extending to radii much greater than R_0. As mentioned already the method which I have just described is the one which has been used to obtain masses for other spiral galaxies.

The Schmidt technique can be generalised by recognising that the density of the Galaxy is not really constant on spheroids of a single eccentricity. Even the visible components of the Galaxy can be divided into a thin disk and a thick disk, the central nuclear bulge and the halo. The spatial distribution of any dark matter is certainly different from that of the visible matter. The discussion of such multicomponent models of the Galaxy, and other galaxies, is too detailed for the present book and I shall just say that such treatments exist.

Do galaxies have massive haloes?

Although it is generally believed that large spiral galaxies do have masses of $10^{11}M_\odot$. or a few times that value, there have recently been suggestions that they might have substantially larger masses. I have already mentioned that a spherical or spheroidal distribution of mass outside the Sun would not exert any net gravitational force in the solar neighbourhood, so that observations of the rotation curve with the solar radius do not prevent the existence of a massive halo. I complete this chapter by explaining why the existence of a massive halo has been suggested and by mentioning observations relevant to its existence. At least four different arguments point to the possibility of galaxies being more massive than is generally believed. The first relates to the hidden mass problem which has been mentioned in Chapter 3 and which will be discussed again in Chapter 8. Estimated masses of individual galaxies in groups or clusters of galaxies are such that, when the Virial Theorem in the form (5.8)

$$M = R\langle v^2\rangle/aG \qquad (5.8)$$

is applied to a cluster, the two sides of the equation are far from being equal. The *virial mass* is much greater than the directly estimated mass. It has therefore been suggested that there must be missing mass in the clusters. This mass may, at least partially, be intergalactic matter but the discrepancy would be eased if galactic masses were larger than is generally believed. For individual spiral galaxies the principal evidence for additional mass is the flat rotation curves at large radii, which have been mentioned in Chapter 3. A spherical distribution of mass with density $\rho = \rho_0 r_0^2/(r_0^2 + r^2)$ produces a flat rotation curve for $r \gg r_0$ and a total mass which increases linearly with r at large values of r. Although a massive halo could not increase the mass of our Galaxy within radius R_0 above the value (5.15) the Galaxy could be significantly more massive if the rotation curve is flat to radii several times R_0.

Another method of estimating the mass of the Galaxy and of the entire Local Group of Galaxies involves studying the motions of the galaxies in the Local Group. There is a problem that only one component of the motion of the other galaxies in the Local Group can be measured, that towards or away from the Galaxy. Mass estimates thus depend on what relation is assumed between the radial and transverse velocities. The next crucial question is whether the Local Group is a group of galaxies which is coming together for the first time or whether it has reached an equilibrium governed by the Virial Theorem. These techniques provide rather uncertain estimates of the total mass of the Galaxy. They all provide a mass of the Galaxy significantly greater than the $1 - 2 \times 10^{11}M_\odot$ obtained from direct studies of the disk and there are suggestions that the total mass of the Local Group could be greater than $3 \times 10^{12}M_\odot$. As most of the mass must be in Andromeda and the Galaxy, this indicates that there could be a very large amount of hidden matter in the Galaxy.

The other two arguments for massive haloes are more subtle and cannot be discussed in detail here, although one will be discussed further in Chapter 7. This latter relates to the problem of the origin of the chemical elements and the chemical evolution of the Galaxy. It is known that very old stars in the Galaxy contain very little in the form of elements heavier than hydrogen and helium and the *big bang* cosmological theory suggests that the Universe may originally have contained no heavy elements. Although the observations of low heavy element abundances in old stars are in qualitative agreement with the cosmological theory, there remains the problem of where the heavy elements in the oldest known stars were produced. Either they were already present when the Galaxy was formed or they were produced in its formation phase. In this case they may have been formed in a massive explosion in the galactic centre or the first period of star formation occurred in a more massive and larger Galaxy than the present halo, disk and nucleus. These parts of our Galaxy would still have been in the form of gas at that stage and would have been enriched by heavy elements produced in the first generation stars. I shall discuss this further in both Chapters 7 and 8.

The other suggestion is highly theoretical. The Galaxy (and other spiral galaxies) is an extremely flattened system and it may be asked whether such a flat configuration can be expected to be stable. If the observed distribution of matter in the disk is disturbed, will it return to its original shape and distribution or will it get more distorted and take up an entirely different shape? A variety of calculations have suggested that very flat galaxies like our own should be unstable but that the instability could disappear if the total mass of the galaxy is very much larger than generally believed and if the galaxy has a massive approximately spherical halo. The argument is at present suggestive rather than conclusive, because the calculations have only shown that the flat disk is unstable to small disturbances and have not predicted the final effect of the instability.

Possible methods of detecting massive haloes

Finally, I must ask whether there are observations which may tell us something about the possible existence of such massive haloes. The first point to be made is that, if it proves quite impossible to detect any gravitational influence from them, as has been suggested, they can only be studied if they are luminous. There exist models of galactic formation and evolution which suggest that most of the massive halo was converted into high mass stars which have long since evolved and are now dead stars. If this is true, we cannot hope to detect them directly and we could have a massive halo largely composed of black holes.

Suppose instead that the first generation of stars had a mass distribution similar to that at present observed in our own and other galaxies. Might we expect to observe these extreme halo objects? In the case of galaxies other than our own, might we be able to observe faint extensions of galactic images to very large distances indicating that large haloes exist? This may prove possible and this would be a direct detection. It is, however, also possible to demonstrate that a galactic halo which was very large indeed could contain a very substantial mass of ordinary stars, which would not produce a surface brightness which would be readily detectable. In the case of our own Galaxy, the situation is somewhat different. If our Galaxy does contain a large halo, the extreme halo stars do not spend all of their time there. Like the high velocity stars of the ordinary halo, they must move in orbits which cause them to pass through the galactic disk period- ically and at any time the solar neighbourhood must contain some extreme halo stars. These stars would have very high velocities and it would make sense to search or such stars in the solar neighbourhood. They should also have negligible heavy element abundance but for this to be studied they will have to be bright enough for a good spectrum to be obtained. Searches for very high velocity extreme halo stars have been made and have proved negative and some astro- nomers feel that there is already clear evidence that our own Galaxy does not contain a massive halo of ordinary stars. As mentioned earlier, that does not rule out the possibility of a halo largely composed of dead stars; extreme halo black holes in the solar neighbourhood would not have been detected!

Another possibility is that the massive haloes of galaxies are not made of ordinary matter but are composed of weakly interacting massive particles such as neutrinos, if they have a finite but small mass, or other particles whose existence has been suggested by theoreticians but which have not yet been detected experimentally. I shall say more about this possibility in Chapter 8. We shall see there that there are arguments that most of the hidden matter in the Universe must be weakly interacting elementary particles. If that is the case, it is difficult to avoid the conclusion that they are an important component of galaxies. Mean- while I should remark that it has been suggested that the halo particles might, in fact, be detected by a terrestrial experiment and several such dark matter experiments have been proposed although none is yet operational. It must be remembered that, although I refer to a massive halo, the density of such halo particles will be higher in the galactic midplane than above it.

Summary of Chapter 5

In this chapter I have described several methods by which the masses of galaxies can be determined. I first discussed the possibility of obtaining the masses of double galaxies by studying their relative motions. I explained that, because of uncertainties in the orientation of the orbits of individual pairs of galaxies only statistical information can be obtained in this way. I then described how the Virial Theorem can be used to obtain masses of individual elliptical galaxies and the total mass of clusters of galaxies. In the latter case the total mass of a cluster is often found to be significantly greater than the sum of the estimated masses of all galaxies in the cluster. This is the so-called virial mass discrepancy.

Most of the chapter has been concerned with the determination of the masses of highly flattened galaxies from their rotation curves, with the discussion being particularly related to the Galaxy. It appears that the assumption that the mass distribution is flattened like the light distribution should give a reasonably accurate mass for the region for which there is a reliable rotation curve. Although a spherical distribution of mass producing the same rotation curve would have a greater mass, the ratio of the two masses would be less than a factor of 2.

The total masses of galaxies are much more uncertain than this. Several independent arguments have suggested that galaxies might have massive low density haloes. These include the existence of the virial mass discrepancy in clusters of galaxies and suggestions that highly flattened disks without massive haloes should be unstable. However, the strongest argument as far as spiral galaxies are concerned is their very flat rotation curves out to radii at which there is very little light. This shows that, even if the mass is flattened like the light distribution, it cannot follow the light distribution in detail. The nature of the dark matter in galaxies remains uncertain but at least part of it may not be ordinary matter but may instead be weakly interacting elementary particles.

6

The interstellar medium in our Galaxy

Introduction

There will be no attempt in this book to describe the detailed physical properties of the interstellar medium (gas, dust, cosmic rays, magnetic field) in our own and other galaxies; that would require a whole book to itself. It is, however, necessary and desirable to discuss *some* of its properties for several reasons. In the first place I believe that galaxies were initially composed of gas alone, so that the process of galactic evolution is largely a process of conversion of gas into stars with the subsequent evolution of the stars. In the second place, if even five per cent of the visible mass of a galaxy is in the form of gas at the present time as we have seen to be true for our Galaxy, it is not really possible to discuss the structure of the galaxy whilst ignoring the gas. The gas content of galaxies will be discussed in this and the following two chapters.

Here I discuss the present structure of the gas disk in our Galaxy, with some comments about other galaxies and with some remarks about the possible past behaviour of the disk. In Chapter 7 I shall discuss the chemical evolution of galaxies, about which information is obtained by a study of the chemical composition of stars of different ages and of the present composition of the interstellar gas. Finally, in Chapter 8, I give a very tentative discussion of the problem of galaxy formation and also of the very late stages of galactic evolution. The order of these three chapters is not really the most logical one. It would be much more logical to start with the formation of galaxies and to work through their past evolution to their present structure. However I have chosen to reverse the order because much more is known about the present state of galaxies than about either their past evolution or their origin.

I have already mentioned in Chapters 2 and 3 that galaxies contain interstellar gas and dust as well as stars and that the gas forms a larger fraction of the mass of spiral and irregular galaxies than of elliptical galaxies. In Chapter 5 some observations of the gas have been used in studying the connection between galactic rotation and galactic masses. In spiral galaxies, as we have seen, most of

Figure 76.   The helical motion of a cosmic ray particle about a line of magnetic induction.

the gas forms a very thin disk comparable in thickness with the disk of the bright stars. I certainly believe that the gas disk, like the star disk, is flattened because it is rotating but, as we shall see shortly, that does not immediately imply that its thickness should be the same as that of the star disk. I therefore wish to discuss what are the factors which determine its precise thickness. To be specific I shall discuss the Galaxy and, in particular, the region of the galactic disk near to the Sun.

Cosmic rays and magnetic fields

Before I do this, I must first mention a further constituent of the Galaxy. We observe in our neighbourhood of the Galaxy the existence of energetic cosmic ray particles. These are few in number but they possess a kinetic energy density (energy per unit volume) which is comparable with the kinetic energy density of the interstellar gas. The gas is much more massive but it is moving very much more slowly. Cosmic ray particles are moving with a speed which is only very slightly less than the speed of light and they would escape from the Galaxy in less than 10^5 years, if there were nothing to prevent their escape. Their free escape is prevented by the interstellar magnetic field. In what follows I shall discuss *magnetic induction (magnetic flux density) B*, rather than *magnetic field strength* and this is typically of value 3×10^{-10} tesla.† Because cosmic ray particles are electrically charged they are constrained to move in helical paths about the lines of magnetic induction (fig. 76) and they can in the first instance be expected to remain in the Galaxy unless the lines of magnetic induction themselves leave the Galaxy.‡

Magnetic fields can only be produced either by magnetic materials or by electric currents, and in the Galaxy it is electric currents which are responsible. As these currents can themselves only be produced by the motion of positive relative to negative electric charges either in the (lightly ionised) interstellar gas or in the cosmic rays themselves, it perhaps sounds surprising that I am discussing the magnetic field as if it is an independent object. Two comments can be made on this

---

† In comparison the Earth's surface magnetic induction is of order 10^{-4} tesla.

‡ The line of magnetic induction through any point can be traced out by following the direction of the magnetic induction vector through that point. In a static field the lines are fixed and because of the absence of magnetic poles they must close or extend to infinity in both directions. Lines of induction can similarly be defined at any instant in a time-varying field. In what follows lines of induction will be called magnetic field lines.

point. The first is that only extremely small relative motions of positive and negative charges are needed to produce the observed magnetic field. A typical interstellar gas density is 10^6 m^{-3} and, even if it is only lightly ionised, the number of charged particles may be in excess of 10^4 m^{-3}. If the magnetic induction B varies significantly over a distance L, it is related to the current density j and the number density n and the relative velocity v of the charged particles by the approximate relation

$$B/L \approx \mu_0 j = \mu_0 nev, \tag{6.1}$$

where μ_0 is the permeability of free space ($4\pi \times 10^{-7}$ H m^{-1}) and e is the charge of an electron. For a magnetic induction of 3×10^{-10} tesla with a scale of variation of 100 pc, which I believe to be characteristic of the large scale magnetic field, the relative speed of negative and positive charges need only be about 5×10^{-8} m s^{-1}. This is quite insignificant compared with both the thermal speeds of interstellar particles and the random speeds of interstellar clouds which are both typically in the range 10^3–10^4 m s^{-1}. The speed will also remain insignificant in regions where the scale of the magnetic field, L, happens to be smaller by many orders of magnitude.

Permanence of interstellar magnetic fields

The second comment is that once a current is flowing and a field has been produced, the field can only be reduced or destroyed if the current carriers collide thus reducing the relative velocity and hence the current. Another way of putting this is to say that currents decay because a medium has a finite electrical resistivity, η. The time of decay, τ_D, of currents which have a scale variation L can then be shown to be given by the approximate relation

$$\tau_D \approx L^2/\eta. \tag{6.2}$$

Because of the large distances involved in the interstellar medium, it turns out that the time τ_D is often much longer than the believed age of the Galaxy. Some values of decay times of cosmic magnetic fields are shown in Table 9. The value quoted for the interstellar medium is too large because, in fact, the electrical resistivity is anisotropic being larger perpendicular to the magnetic field than along it. This does not change the conclusion that in the case of the interstellar medium, a large scale magnetic field will only decay in any time of interest if the medium's degree of ionisation is very low. Such conditions are most likely in dense gas clouds, which are in any case rather short-lived. In contrast magnetic fields in stars may or may not be able to survive for their entire lifetime and the Earth's magnetic field cannot possibly have been freely decaying for the past 4.5×10^9 yr. It must be continuously regenerated.

Interaction of a magnetic field and fluid motions

If a magnetic field is present in a conducting fluid (liquid or gas) and if its time of decay is long, the field can be regarded as a permanent entity. This does

Table 9. *Approximate decay times of cosmic magnetic fields.*
The decay times depend strongly on the degree of ionisation of
the material. The stellar interior and the hot interstellar gas are
essentially fully ionised. The substantial difference between
decay times for the cold interstellar gas and the gas cloud
arises because it is assumed that cosmic rays keep the former
at a much higher level of ionisation than would be expected
for its temperature. The figure for the Earth relates to a liquid
metallic core.

Object	L/m	T/K	n/m^{-3}	τ_D/yr
Stellar interior	10^8	10^6	10^{30}	2×10^9
Hot interstellar gas	10^{17}	10^4	10^4	2×10^{24}
Cold interstellar gas	10^{18}	10^2	10^6	10^{13}
Cold gas cloud	10^{17}	10^2	10^8	3×10^8
Earth	10^6			2×10^4

not, however, mean that no change can occur in the value of the magnetic
induction. The reason for this is that the magnetic field lines are *tied* to the
conducting fluid and if the fluid moves, they must also move. The quantity which
is conserved in such a motion of a highly conducting fluid is the *flux* of magnetic
induction. This concept is illustrated in fig. 77. The flux through a small area A
bounded by a set of fluid particles is BA. If the particles move in such a way as to
reduce A, B increases, and vice versa.† If fluid motions reduce the scale of the
field sufficiently, some decay may in fact occur. If the increase in the field
produced by fluid motions just balances the decay, the field is said to be dynamo
maintained.

The forces acting in the galactic disk

After this brief digression on the properties of magnetic fields in conducting fluids, I now turn specifically to the interstellar medium, regarding the
interstellar magnetic field as something that has an almost independent existence.
In what follows it will be sufficient if it neither grows nor decays in a few times
10^7yr because that is a characteristic time of gas motions in the disk. As mentioned
already, the cosmic rays can only be prevented from leaving the interstellar
medium at the speed of light by the restraining influence of the magnetic field.
They in turn, acting as a highly conducting fluid in the way which I have just
described, try to pull the magnetic field lines out of the Galaxy with them.
However, the magnetic field is also embedded in the interstellar gas and is

† These ideas concerning the interaction of conducting fluids and magnetic fields were first derived
theoretically but they were soon verified by experiments involving mercury. It is very difficult to stir
mercury placed between the poles of a powerful electromagnet, but if the fluid is moved the
magnetic induction changes in the manner predicted by the theory.

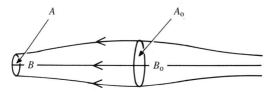

Figure 77.   The conservation of magnetic flux. The arrows are along magnetic field lines. $B_0 A_0 = BA$.

produced by currents flowing in the gas, and it cannot leave the Galaxy without taking the gas with it. If a sufficiently large quantity of cosmic rays were suddenly produced in the Galaxy, this could happen and something very much like this may have happened in some of the other explosions in other galaxies, which have been mentioned in Chapter 3. In particular some theoretical models of radio galaxies require a large input of relativistic charged particles in their central regions. These particles subsequently escape in two jets which feed the observed radio lobes.

In the galactic disk there is one other important force. The gas is attracted to the galactic midplane by the gravitational field of both the stars and the gas itself. If there were no magnetic field and cosmic rays, the thickness of the gas disk would be determined by a balance between the random velocities of the gas and the gravitational attraction of the stars and the gas, just as the thickness of the star disk is determined, as described in Chapter 4, by a balance between the random velocities of the stars and their mutual gravitational interactions. As the cosmic rays are coupled to the gas through the magnetic field, they try to pull the gas away from the galactic midplane and the disk is thicker as a result. There is also one other factor which may possibly be important. It is not clear whether or not there is any significant amount of intergalactic matter near to the Galaxy. If there is, it may exert a pressure on the outer boundaries of the disk and this will have the effect of making the disk thinner than it would otherwise be. There are some observations which suggest that the mass of the disk may still be increasing as a result of infall of intergalactic gas, as will be mentioned further in the next chapter.

The evolution of the galactic disk

At any time in galactic history, I can use the balance of the above mentioned forces as a basis for a discussion of the gas disk and I shall do this shortly for the case of the solar neighbourhood at the present time. First I shall make a few remarks about what happens to the balance of forces as the Galaxy evolves and there is an exchange of identity between different components of the disk. For example stars are formed out of the interstellar medium and there is also a variety of processes, explosive or otherwise, by which existing stars lose mass which is returned to the interstellar medium. As matter ejected by stars tends to have considerably more kinetic energy per unit mass than the general interstellar gas, the net effect of star formation and mass loss from stars is to increase the

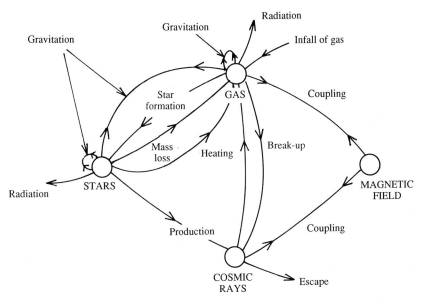

Figure 78.   Physical processes connecting the different components of the Galaxy.

kinetic energy per unit mass of the interstellar gas. This alone would lead to an increase in the thickness of the gas disk. The increase is not as great as might be expected because studies of the dynamics of matter expelled in a supernova explosion show that most of the energy is radiated and less than ten per cent contributes to the kinetic energy of the interstellar gas. There are, in addition, many competing effects and these are shown schematically in fig. 78. A major cause of loss of energy arises in cloud-cloud collisions. The way in which the thickness of the galactic disk changes with time depends on the interaction of all of these different factors.

Observations of the distribution of stars with height above the galactic plane which are described in Chapter 2 (Table 3 on page 26), might give some information about the thickness of the gas disk when they were formed. I could then compare this information with theoretical ideas about how the disk thickness should have changed. The situation is, however, not completely clear. Although stars rarely collide with other stars, they do more often collide with interstellar clouds which are much larger and more massive. The typical result of such a collision will be to transfer kinetic energy from the more massive cloud to the less massive star at least until the average kinetic energy of individual stars and clouds is similar. A similar effect might be produced by transient changes in the gravitational field acting on a star as it passes through spiral arms. This means that old stars *might* occupy a thicker disk than they did when they were formed and this is generally believed to be true. At present there is no clear indication that the thickness of the disk has been changed significantly since most of the mass of the Galaxy has been in the form of stars.

Equilibrium of the gas, cosmic ray, magnetic field system

I now turn to some more quantitative ideas about the structure of the disk of the Galaxy. Because the effective thickness of the galactic disk near the Sun is no more than a few per cent of its radius, its properties vary much more slowly in the radial direction, than in the direction perpendicular to the disk. This means that it is possible, to a first approximation, to regard the disk as a plane parallel system which is infinitely long in the direction perpendicular to its thickness. I do not therefore consider any further the $\bar{\omega}$ component of the equation of equilibrium of the disk, which essentially relates the $\bar{\omega}$ component of the gravitational field to the circular velocity in the manner which I have discussed in Chapter 5. Instead I study only the z component of the equation of equilibrium in a frame of reference which is rotating with the local circular velocity, $v_{\phi 0}$. Although the Galaxy is rotating differentially as I have discussed in Chapter 2, we have seen that the time taken for a star or gas cloud to move up and down through the disk is several times less than the period of galactic rotation. This means that I should be able to ignore the differential rotation to a first approximation when discussing the disk structure in the z direction. Gas clouds do, of course, have an epicyclic motion in the radial direction similar to that of the stars, but they are likely to suffer collisions and possibly loss of identity in a time short compared to that of galactic rotation.

Now let me study a small element of matter in such an infinite disk and consider the balance of forces acting on it assuming that it is in equilibrium. I am here considering the system of *gas*, *cosmic rays* and *magnetic field* and I treat the stars only as a source of gravitational attraction for the gas. An element of matter of thickness δz and surface area δS situated at height z above the galactic midplane (fig. 79) is acted on by a gravitational force $\rho \delta S \delta z g_z$ where ρ is the mass density of the gas and the cosmic rays and $|g_z|$ is the component of the gravitational field acting towards the plane of the Galaxy. Now, if E_{cr} is the energy density of cosmic rays, including their rest mass energy, which is negligible in comparison with their kinetic energy, the mass density of cosmic rays is E_{cr}/c^2, using Einstein's mass/energy relation. As we shall see shortly, the observed value of E_{cr} is very similar to $\rho_{gas}\langle v_{gas}^2 \rangle$, where ρ_{gas} is the mass density of the gas and $\langle v_{gas}^2 \rangle$ is its mean square speed. As $\langle v_{gas}^2 \rangle \ll c^2$, $\rho_{cr} \ll \rho_{gas}$, and it can be neglected to a good approximation. Thus the gravitational force on the element is

$$\rho_{gas} g_z \delta z \delta S. \tag{6.3}$$

Now the gas and cosmic rays can be regarded as fluids which exert a pressure on the upper and lower surfaces of the element. The net downward force due to these two components is

$$[P_{gas}(z + \delta z) + P_{cr}(z + \delta z) - P_{gas}(z) - P_{cr}(z)]\delta S, \tag{6.4}$$

where P_{gas} and P_{cr} are the pressures exerted by the gas and cosmic rays respectively. In introducing P_{gas} and P_{cr} I am in effect assuming that the random velocities of the gas have an equal amplitude in all directions and that the magnetic

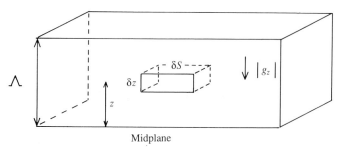

Figure 79. A small element of the galactic disk.

field makes the cosmic ray motions isotropic. Neither of these assumptions will be precisely true and as a result the pressure exerted by both gas and cosmic rays will depend slightly on the direction in which it is acting. The force per unit area exerted by a magnetic field is in general more complicated than a simple pressure. There is a pressure component but in addition there is a tension in the field lines which tries to straighten bent field lines. This leads to the propagation of hydromagnetic waves along the field lines, which are similar to sound waves on a stretched string. The force can be closely approximated by a pressure, P_{mag}, when the field is purely horizontal and when the field varies in the z direction very much more rapidly than it varies in the other directions. Although the galactic magnetic field is not *purely* horizontal, it appears from interpretations of observations, such as those of interstellar polarisation described on page 48, that the field is mainly parallel to the galactic plane. It is probably therefore reasonable to add one term

$$[P_{mag}(z + \delta z) - P_{mag}(z)]\delta S \tag{6.5}$$

to (6.4) to take account of the force exerted by the magnetic field. Equating the net force due to the pressures to the gravitational force, I obtain

$$\frac{dP}{dz} \equiv \frac{dP_{gas}}{dz} + \frac{dP_{cr}}{dz} + \frac{dP_{mag}}{dz} = \rho_{gas}g_z \tag{6.6}$$

where g_z is negative because the gravitational field (z component) acts towards the midplane.

Before discussing this equation any further I ought to satisfy myself that it is reasonable to suppose that the galactic disk is in equilibrium. In general terms it is easy to see that this is likely to be true. I have already seen in Chapter 4 that matter would fall from the top of the galactic disk to the midplane in about 2×10^7 yr if there were nothing to prevent it. This means that, if the gravitational force and the pressure gradient were not in close balance, I should expect the disk to change its thickness significantly in such a time. This time is very short compared to all other times important in galactic evolution; the period of galactic rotation, the evolution time of most stars and, hence, the timescale of chemical evolution of the Galaxy. It seems clear that to a first approximation the galactic disk *is* in equilibrium. It must, however, be repeated that individual stars and gas clouds which help to make up the disk of the Galaxy may oscillate relatively freely through the disk. The system is in equilibrium in the sense that although the individual components are oscillating, they are distributed in such a way as to maintain a steady state.

Properties of the interstellar medium near to the Sun

I now return to equation (6.6). I cannot solve this equation in any formal sense but I can try to demonstrate that the observed values of the quantities which enter into the equation are consistent with the disk being in equilibrium. As there are considerable uncertainties in the values of all of the quantities there is no point in trying to treat the equation too precisely. First, let me consider the forms of the different terms in equation (6.6). The gas in the Galaxy is not smoothly distributed in the disk. Apart from concentration in the region of the spiral arms, which I shall discuss further on page 144, the galactic gas is partly in the form of clouds of largely un-ionised gas at a temperature of order 50 K to 100 K and partly in the form of hotter and partially ionised gas with a much lower density and a temperature of about 10^4 K. There is also even lower density gas at about 10^6 K, which occupies a significant fraction of the volume of the disk. Finally there are dense molecular clouds which are few in number but individually very massive ($\sim 10^5 M_\odot$) and with a temperature which can be as low as 10 K.

The neutral gas is the most readily observable by use of the 21 cm radiation of hydrogen as described in Chapter 2, but it is obvious that it cannot be all of the gas that there is. A typical cold cloud is observed to be of such a mass that its internal gravitational forces are insufficiently strong to hold it together against its internal pressure. Such a cloud would expand freely and disperse if there were not a warm or hot intercloud medium exerting a comparable pressure. If the thermal pressures of the cold, warm and hot components of the gas are all about the same, their densities must very roughly be in inverse proportion to their temperatures and it is then clear that most of the mass of the interstellar medium is in the cold clouds. These clouds are moving through the disk with random velocities which are about a factor ten higher than their internal thermal velocities. I can then write

$$P_{\text{gas}} = \tfrac{1}{3}\rho_{\text{gas}}\langle v_{\text{gas}}^2\rangle, \tag{6.7}$$

where ρ_{gas} and $\langle v_{\text{gas}}^2\rangle$ have been defined earlier and where $\langle v_{\text{gas}}^2\rangle$ can be taken to be the random velocities of the cold clouds, as their motions make the largest contribution to the kinetic energy and hence to the pressure.

Direct observations of cosmic rays are restricted to the Earth and its immediate surroundings, although radio emission from other parts of the Galaxy is believed to be produced by cosmic ray electrons spiralling in the magnetic field, as has been mentioned briefly on page 48. When account has been taken of the effect on their motion of the magnetic fields of the Sun and Earth, it appears that cosmic ray particles are reaching the Earth equally from all directions in space so that they clearly do not have a local origin. It is believed that they originate throughout the Galaxy and most theories associate their production in one way or another with the explosions of supernovae. It seems reasonable to suppose that their number density near the Earth is characteristic of other regions in the middle of the galactic disk. In any case, because the cosmic ray particles are moving with relativistic speed, I can write

$$P_{cr} = \tfrac{1}{3}E_{cr}, \tag{6.8}$$

a relation which is true for all relativistic systems. Finally, if the force per unit area exerted by the magnetic field can be approximated by a pressure, this has the form

$$P_{mag} = B^2/2\mu_0, \tag{6.9}$$

where, as before, B is the magnitude of the magnetic induction and μ_0 the permeability of free space.

If I suppose that the half thickness of the galactic disk is Λ (fig. 79) and that there is no external pressure on the top of the disk, I can approximate the pressure gradient on the left hand side of equation (6.6) by the pressure difference between the top and bottom of the disk divided by Λ, that is by

$$-[P_{gas} + P_{cr} + P_{mag}]_{z=0}/\Lambda. \tag{6.10}$$

This should be a first approximation to the pressure gradient half-way between the midplane and the top of the disk and I can equate it to the gravitational force evaluated at the same point. Thus

$$-[P_{gas} + P_{cr} + P_{mag}]_{z=0}/\Lambda = [\rho_{gas}g_z]_{z=\Lambda/2}. \tag{6.11}$$

I have approximate values of all of the quantities in this equation from observation. I can insert these into the two sides of equation (6.11) and see whether they turn out to be approximately equal.

Oort's limit

Before I do this, I return to a discussion which was left incomplete at the end of Chapter 4. In Chapter 4 I used observations of stellar velocities to obtain values of g_z throughout the galactic disk and I commented that once g_z was known a value could also be obtained for the total density of gravitating matter. With the assumption that gradients in the z direction are very much larger than gradients in the $\bar{\omega}$ direction, Poisson's equation for the gravitational potential Φ takes the simplified form

$$d^2\Phi/dz^2 = -4\pi G\rho, \tag{6.12}$$

where ρ now includes the mass density of stars and any other matter. Near the galactic midplane I found

$$g_z = -\lambda z, \tag{6.13}$$

with $\lambda \approx 10^{-29}\,\mathrm{s}^{-2}$. Noting that $g_z = d\Phi/dz$, (6.12) and (6.13) can be combined to give a value for ρ near to the midplane, which is

$$\rho \approx 0.15 M_\odot\,\mathrm{pc}^{-3}, \tag{6.14}$$

which I have already used in Chapter 5. This value of the density of matter near the galactic midplane is known as *Oort's limit*.

The interest in this value is that it can be compared with the known density of matter in the same region. If the known density is less than Oort's limit, it implies that there is *hidden matter* in the solar neighbourhood. If the known density were greater, this would, in contrast, imply excess matter and cause serious embarrassment. In fact, observations have always led to a value less than Oort's limit. At the time of the original discussion, only about half of the density could be accounted for. Since then further faint stars and gas (particularly atomic and molecular hydrogen) have been detected and the typical values found by recent investigations were

$$\rho_{\text{stars}} \approx 0.064 M_\odot \, \text{pc}^{-3}. \quad \rho_{\text{gas}} \approx 0.024 M_\odot \, \text{pc}^{-3}. \tag{6.15}$$

Different investigators have found a variety of values for the local density and are also in disagreement about whether there is any further undetected matter, as has already been mentioned in Chapter 5. I will regard it as possible but not certain that there is a substantial amount of local hidden matter. Any remaining hidden matter may be very faint or dead stars or it could be further undetected gas. Alternatively it could be weakly interacting elementary particles. It must be noted that the relative quantity of matter in gas and stars implied by (6.15) does not contradict earlier statements about the relative masses of stars and gas in the Galaxy as a whole. The relative amount of gas is much lower in the nuclear regions which contain much of the mass of the Galaxy and the proportion of gas is higher than the average near the Sun.

I now return to equation (6.11) and consider the observed values of all the quantities contained in it. I have already stated that the magnetic induction has a value of order 3×10^{-10} tesla, although it is certainly not uniform. Using equation (6.9) this gives

$$P_{\text{mag}} \approx 4 \times 10^{-14} \, \text{N m}^{-2}. \tag{6.16}$$

The energy density of cosmic rays near to the Earth is approximately $10^6 \, \text{eV m}^{-3}$ or $1.6 \times 10^{-13} \, \text{J m}^{-3}$. This gives

$$P_{\text{cr}} \approx 5 \times 10^{-14} \, \text{N m}^{-2}, \tag{6.17}$$

showing that the pressures of the magnetic field and cosmic rays are approximately equal. This will be discussed further below. Finally the random velocities of the gas are about 10 km s^{-1}, which combined with (6.15) gives

$$P_{\text{gas}} \approx 6 \times 10^{-14} \, \text{N m}^{-2}, \tag{6.18}$$

which is again very little different from the other two pressures. They are certainly all equal within the accuracy to which they are known. The half thickness of the disk is about 100 pc or 3×10^{18} m so that I can estimate the left hand side of (6.11) to be $5 \times 10^{-32} \, \text{N m}^{-3}$. On the right hand side, it is clear from equations (6.12) and (6.13) that g_z varies with height much more rapidly than does ρ_{gas}; if g_z were exactly linearly dependent on z, ρ_{gas} would be constant. I shall not make much error by using the midplane value of ρ_{gas} and by evaluating g_z at 50 pc. If this is done, the right hand side of (6.11) is approximately 3×10^{-32} N m^{-3}. Although this does not represent exact equality, it is a remarkably good agreement considering that I have treated the equation very crudely and that none of the quantities are known very accurately. Although a calculation of this type cannot *prove* that the disk is in equilibrium, it certainly gives us no reason to depart from our tentative conclusion on page 135 that the disk is in equilibrium.

I have already made a few comments on page 133 about possible changes in the thickness of the galactic disk with time. I am now in a position to make a few more quantitative comments using equation (6.11) in the equivalent form

$$\Lambda \approx [\langle v_{gas}^2 \rangle/3 + (P_{cr} + P_{mag})/\rho_{gas}]_{z=0}/|g_z|_{\Lambda/2}. \tag{6.19}$$

From this equation it is easy to see that, since g_z will not change if mass is neither lost from nor gained by the Galaxy, either an increase in $\langle v_{gas}^2 \rangle$ or a decrease in ρ_{gas} leads to an increase in the thickness of the disk, if the properties of the magnetic field and cosmic rays do not also change. Conversely a decrease in $\langle v_{gas}^2 \rangle$, which may arise because the gas cools or because collisions between gas clouds and stars give energy to the stars, will cause the gas to get thinner. The role of the cosmic rays or the magnetic field acting alone is quite obvious. If all of the quantities change simultaneously, the result is more complicated and such has certainly been the case in past galactic history.

The galactic halo and intergalactic matter

One final remark should be made about our discussion of the equilibrium of the disk. In replacing equation (6.6) by (6.11), I assumed that the pressure at the top of the galactic disk was very low. This may not be true. The Galaxy might be surrounded by an intergalactic medium which exerts a significant pressure and the present observations of the halo of the Galaxy do not exclude the possibility that it contains a gas of very low density but high temperature ($\sim 10^6$ K), whose pressure at $z = \Lambda$ should be included in (6.11). There are also observations of radio emission which suggest that there are both cosmic rays and magnetic fields outside the thin gas disk and these will be mentioned again later. The discrepancy between the two sides of equation (6.11), which is apparent in my discussion, could be partially resolved by the presence of a hot halo, as it would act to decrease the left hand side as the pressure at $z = \Lambda$ would have to be subtracted, but as I have already stated it is not obviously necessary. I shall also explain in the next chapter that it is possible that the galactic disk is still growing in mass by accretion of intergalactic matter.

The approximate equality of gas, cosmic ray and magnetic pressures

I now make some brief qualitative remarks on the approximate equality between the gas, cosmic ray and magnetic pressures. Am I to regard this approximate equality as purely accidental or am I to read something deeper into it? Consider first the cosmic rays. These are confined to the Galaxy by the magnetic field. Is this confinement perfect? It is believed that cosmic rays are continually being produced, probably in connection with the explosion of *super-novae* or by continuing activity in *pulsars* produced in supernova explosions. Once a cosmic ray particle has been accelerated to high energy it will move round the Galaxy occasionally colliding with the interstellar gas particles. As a result of such collisions, atomic nuclei in the cosmic rays can either be slowed down or be broken up into less massive nuclei. The most important process is the conversion of massive nuclei into light nuclei and the observed chemical composition of cosmic rays gives information about the amount of break-up that has occurred and hence about the average age of the cosmic ray particles. From investigations of this type the age of cosmic rays in the solar locality is found to be less than 10^7 yr, which suggests that cosmic rays are not trapped in the galaxy permanently.

The escape of cosmic rays

If the cosmic rays do escape, I must find a reason for this. Although a complete theory has not been worked out, a general idea is as follows. The cosmic rays are supposed to be confined in the Galaxy by the magnetic field. Such a confinement mechanism probably makes perfect sense so long as the energy density of the cosmic rays is significantly less than the energy density of the magnetic field. In such circumstances the magnetic field can act as a rigid containing vessel which the cosmic rays cannot distort. The same is probably not true if the cosmic ray energy density becomes too large; if the cosmic ray energy is too large I might expect the magnetic field container to burst or at least to become leaky.

A somewhat similar problem has been met in experiments designed to produce controlled thermonuclear fusion reactions in a laboratory. The initial idea was that a very hot ionised gas would be confined by a magnetic field, so that it did not come into contact with the walls of the containing vessel. It was not difficult to design *magnetic bottles* which were equilibrium configurations of hot gas contained by a magnetic field but most of these were found to be very unstable when an experiment was performed. This was particularly true if the pressure of the contained gas was comparable with the pressure of the magnetic field. It seems quite possible that the cosmic ray/magnetic field system in the Galaxy suffers from similar instabilities particularly if energy density of the cosmic rays is locally comparable with the energy density of the magnetic field. This *might* account for the observed approximate equality of the magnetic and cosmic ray pressures; if all of the cosmic rays which have been produced in the galactic lifetime were still present, the cosmic ray pressure would be several orders of magnitude higher than the magnetic pressure, but in fact the cosmic rays escape if their pressure becomes too high. If instability is the cause of the escape of cosmic rays, it seems probable that they diffuse outwards from the centre of the disk through a series of fairly local instabilities. If this is true the mean age of cosmic rays near the edge of the disk may be significantly greater than for those near to the Sun, but we have no direct information on this point as we can only observe the chemical composition

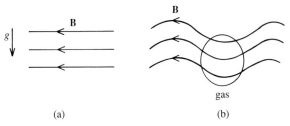

Figure 80. A simple instability of a magnetic field/gas system.

of cosmic rays which reach the Earth. If cosmic rays do escape into the halo as a result of instability, this may account for the radio emission from a region thicker than the main gas disk, which has been mentioned on page 139, because relativistic charged particles moving in a magnetic field radiate what is known as synchrotron radiation.

A simple instability of a magnetic field, gas, cosmic ray system

To make the above ideas slightly more definite I consider one instability which is quite easy to describe and visualise. Consider some horizontal magnetic field lines as shown in fig. 80(a). Associated with the field there will be both interstellar gas and cosmic rays. Suppose now that the field lines are slightly distorted as shown in fig. 80(b). As a result of this disturbance, there will be a tendency for the interstellar gas, which is acted on by gravitational forces, to fall to the low points in the field. In contrast, the cosmic ray particles are unaffected by the gravitational field and they will continue to be approximately uniformly distributed along the magnetic field line. As a result they will exert a greater upward influence on the higher points of the magnetic field lines where the quantity of gas has been reduced. This means that the initial perturbation is likely to grow. Detailed calculations confirm the existence of this type of instability, which may play a part in the formation of interstellar clouds as well as in the behaviour of cosmic rays. It should, however, also be mentioned that there are other more complicated instabilities which I cannot describe here and which are probably more effective in enabling the cosmic rays to escape from the region of the disk occupied by the gas.

Interactions between the gas and magnetic field

I now turn to a discussion of the approximate equality between the gas pressure and the magnetic pressure. Again it is only possible to present a very schematic discussion both because a detailed discussion would be very complicated and because a satisfactory theory has not been worked out. Early in this chapter I mentioned that the galactic magnetic field is unlikely to decay but that its strength could be affected by fluid motions; specifically fluid motions will affect the magnitude of the magnetic induction but will leave the magnetic flux

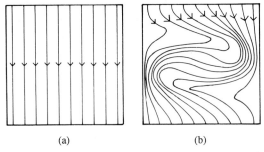

(a)   (b)

Figure 81.   Effect of fluid motions on a magnetic field. A simple vortex type motion moves the straight magnetic field lines (marked with arrows) of (a) into those shown in (b). Where the field lines are closer together the magnetic induction is increased substantially and the average value of the magnetic induction is also increased.

unchanged. Although fluid motions can in principle lead to either an increase or a decrease in magnetic induction (as is obvious by considering a particular set of motions and by asking what happens if they are reversed), if the field has an initially simple structure most fluid motions will tend to increase the induction (see fig. 81). It is therefore quite probable that the average magnitude of the galactic magnetic induction has been increased by the motions of the galactic gas in past galactic history. If the fluid motions are prescribed, this process can continue indefinitely or at least until the motions have distorted the field so much that its characteristic scale of variation L has become small enough for the decay time τ_D given by equation (6.2) to be smaller than other times of importance in galactic evolution. There is, however, another important effect in a system of the type which I am describing. If the strength of the magnetic induction increases, the magnetic forces may become sufficiently strong that they influence the fluid motions.

This has led to the idea that the fluid can distort the magnetic field so long as the energy in the motions is significantly greater than the energy possessed by the field but that the field will resist further distortion and increase in strength once the two energies are comparable. This might account for the approximate equality of the gas and magnetic pressures in the disk given that gas clouds can have moved through the disk very many times during the galactic lifetime. This argument for an *equipartition of energy* between the fluid motions and the magnetic field is not very firmly based and there exists a tendency for it to be used in many problems in astrophysics including some when it is certainly not justified. There are, for example, problems in which the fluid motions have less energy than the field but in which they can still distort the field. These are problems in which there is a mechanism driving the fluid motions which possesses much more energy than either the fluid or the field and which ensures that the field cannot prevent the motion of the fluid.

This discussion also relates to the question of the origin of the galactic magnetic field. Was there a magnetic field in the galactic disk when it was first formed and, if so, was it comparable in strength with the present field or very much weaker?

Attempts to identify a general intergalactic magnetic field have not yet been successful. The present upper limit to its value could have led through conservation of magnetic flux to an initial value of the disk magnetic induction comparable its present value. If the intergalactic field is much weaker, the Galaxy may have started with a correspondingly low field, which has been amplified to its present value as a result of fluid motions. The strength of the initial field could be important for the process of star formation early in galactic history. Because the magnetic energy density is comparable with the kinetic energy density at the present time, it seems clear that the magnetic field must influence the process of star formation. If the magnetic field was not significant in the very early stages of galactic evolution, the distribution of stellar masses, for example, may have been different then.

It must be stressed that the two discussions which have just been given about the interactions of the cosmic rays, magnetic field and interstellar gas should eventually be combined into one simultaneous discussion of the three components of the interstellar medium.

Star formation

I close this chapter with a few remarks on another difficult topic, star formation. The lack of a good understanding of star formation, and of the manner in which it varies in galaxies of different types and during the life of any one galaxy, is perhaps the major problem in astronomy today. Developments in infrared and millimetre astronomy are providing increasing observational information about star formation in our own and nearby galaxies today but it is not easy to see how comparable information can be obtained about star formation in early galactic evolution. In order for interstellar gas to be converted into stars, the matter must increase its mean density by a factor of about 10^{24}. The collapse of a region of interstellar gas which may lead to star formation is likely to occur if it reaches a state in which the magnitude of its gravitational potential energy is greater than its kinetic energy. Typically the interstellar medium does not appear to be in a state of incipient gravitational collapse; if it were, star formation would be extremely efficient and there would probably be no interstellar gas left by now. Star formation apparently *was* very efficient when some galaxies first formed. Whether or not this was the case in our Galaxy is unclear as will be discussed in the next chapter but it is believed to be this early efficiency of star formation which has left elliptical galaxies with little gas now. In a spiral galaxy like our own in which a significant fraction of the mass remained in gas after the formation phase, star formation may now occur if some process increases the mean density and/or reduces the mean random velocity of a region of the interstellar gas. In fact, an increase of density will probably also be accompanied by a reduction of random kinetic energy because cooling processes become more efficient when the density increases.

One process that has been suggested involves the collision of the interstellar clouds which we observe. These are not gravitationally bound and only exist for any significant time because they are surrounded by a hotter intercloud medium which prevents them from expanding. If interstellar clouds collide they may coalesce and eventually this process may produce clouds that are massive enough for gravitational forces to become dominant. Such clouds are the giant molecular clouds, which have been mentioned several times already, and they are thought to be an important site of star formation. In addition to cloud/cloud collisions several other methods of compressing gas to form stars have been suggested.

Spiral density waves and star formation

In Chapter 3 I have introduced the idea that the spiral structure of a galaxy is associated with a wave motion so that the material in a spiral arm is not the same at all times. This idea arises because all parts of a galaxy do not rotate with the same angular velocity and a spiral structure which was made up of a fixed set of particles would be distorted out of recognition after a few rotation periods. It is therefore believed that a spiral wave pattern moves through a galaxy in such a way that the overall appearance of the galaxy remains much the same for many rotation periods, although the matter in the spiral arms is changing continuously.

According to this picture, as the spiral pattern rotates the gas in any region of the Galaxy experiences an increase in density as the spiral wave passes through it. This process may cause some of the gas to enter into gravitational collapse but it is unlikely to lead to the direct formation of individual stars as can be seen by a simple order of magnitude argument. Gravitational collapse of a mass of gas of radius R and mean square random velocity $\langle v^2 \rangle$ can occur if $GM^2/R \gtrsim M\langle v^2 \rangle$. It is more convenient to eliminate R and to write this inequality in terms of the density ρ. If I ignore numerical factors which do not differ too much from unity, the inequality can be written

$$M \gtrsim \langle v^2 \rangle^{3/2}/G^{3/2}\rho^{1/2}. \tag{6.20}$$

For normal interstellar densities and random velocities inside clouds, this gives $M \gtrsim 10^3 M_\odot$. It is unlikely that the increase in density and/or decrease in temperature in a spiral wave will reduce this value to typical stellar masses which are $M_\odot$. However, once the collapse of a large cloud has commenced so that ρ increases, it can be seen from (6.20) that the mass which can be gravitationally bound decreases. It is believed that the result is that the collapsing cloud breaks up into smaller collapsing masses until eventually a large number of objects of stellar mass are produced. Thus star formation is believed to occur through a process of *successive fragmentation,* and most stars may have been formed in clusters rather than individually.

The evidence that star formation is associated with spiral structure and that this is a wave pattern is qualitatively very good. In the first instance massive main sequence stars which are unambiguously young are mainly situated in the spiral regions both in our own Galaxy and in other spiral galaxies. This observation is most easily made for nearby bright spirals where the whole spiral structure can be observed. We have no reason to believe that high and low mass stars are produced by distinct processes. If they are not, the observation that young massive stars are concentrated in spiral arms indicates that the spiral arms are the main site of star formation. I have explained that the spiral wave is believed to move around the galaxy so that star formation can occur at all parts of the galactic disk in turn. The most massive stars have such a short lifetime that, by the time they have ceased to be luminous, the spiral pattern has hardly changed its position. They should therefore only be found in the spiral regions. Slightly less massive stars are luminous for a fraction of the rotation period of the spiral waves. Stars of this type should therefore be visible not only in the spiral region but also in regions just behind the spiral pattern which were fairly recent sites of star formation. Stars of lower mass, such as the Sun, have a main sequence lifetime which is equal to many rotation periods of the pattern. Such stars should therefore be observed throughout the disk with no significant correlation with the present position of the spiral arms. The observations of the distribution of stars of different spectral types and hence lifetimes are in good general agreement with these predictions, especially when it is recognised that the compression of gas in spiral waves is probably not the only cause of star formation.

It has to be recognised that there are difficulties with the picture which has just been described. Although theoretical discussions of the structure of stellar and gaseous disks have shown that they can sustain spiral wave patterns, it has not been demonstrated that they would automatically grow and then stabilise at the amplitude of spiral structure which is observed. In addition, numerical calculations of the evolution of the structure of self-gravitating disks have shown the appearance of spiral structure but this has frequently been found to be transitory rather than long-lived. Although these investigations are themselves not conclusive, they have led to suggestions that the observed spiral structure is a consequence of star formation rather than the cause of star formation. The idea of *stochastic self-propagating star formation* is that much of star formation is induced by the massive stars of the immediately previous generation. The idea is that supernova explosions and the expanding HII regions around massive stars can compress interstellar gas and promote the formation of new stars. In addition star formation is inhibited where it has just occurred because supernovae raise the remaining gas to a very high temperature. There is also some spontaneous star formation which might, for example, arise from the process of cloud collision and coalescence already mentioned. In this picture the observed spiral structure is supposed to arise from the differential rotation of the Galaxy shearing the distribution of luminous stars into a spiral pattern. Although this process can produce spirals, they tend to have short ragged arms as some spirals do, and it is difficult to see how the *grand design spirals* such as M51 can be produced. It is clear that the overall process of star formation is not yet fully understood.

Further problems arise in understanding exactly how a cloud which enters into gravitational collapse turns into stars and what spectrum of stellar masses are produced. Apart from the fact that it is dificult to follow in a numerical calculation the change in density by about 24 orders of magnitude, particularly if there are significant departures from symmetry, there are also important non-gravitational forces to be taken into account. I have already mentioned the magnetic field. The other is the influence of rotation. Even if the initial cloud is quite slowly rotating, it can be very highly flattened as a result of conservation of angular momentum as it contracts. At first sight, it might be expected that stars would be more rapidly rotating and strongly magnetic than is generally observed. Although there are ways in which both the angular momentum problem and the magnetic field problem can be solved, theory is not yet in a position to predict a spectrum of stellar masses to be compared with observation.

Summary of Chapter 6

This chapter has been concerned with the dynamical properties of those components of galaxies other than stars and in particular with the structure of the disk of our Galaxy near to the Sun. It is demonstrated that the electrical resistivity of the interstellar gas is so low that the galactic field is unlikely to decay and it can be regarded as an essentially permanent component of the Galaxy. The magnetic field is strongly coupled to both of the other main components of the interstellar medium, the gas and the cosmic rays. Electrically charged particles, whether they are cosmic rays or the ionised component of the interstellar gas, are constrained to move in helical orbits around the magnetic field lines. The cosmic ray particles cannot be held in the Galaxy by the gravitational field of the stars and the gas but they are held in by the magnetic field. In turn they try to pull the magnetic field out of the Galaxy. The magnetic field is coupled to the gas which is acted on by the gravitational field of the stars. The thickness of the galactic disk must depend on a balance between these various forces and it is shown that the observed thickness of the galactic disk near to the Sun is consistent with this picture.

This discussion of the equilibrium of the galactic disk could be carried through for any values of the energy densities of gas motions, magnetic field and cosmic rays but observations indicate that near to the Sun all three are approximately equal. It is argued that this approximate equality is unlikely to be purely accidental. It is probable that the magnetic field cannot prevent the escape of cosmic rays if their energy density is locally larger than the magnetic energy density. If cosmic rays are continually produced, their density may then increase only until the approximate equality is reached. As the interstellar gas clouds move through the disk they distort the magnetic field and such distortion may increase the average value of the magnetic induction. However, the magnetic field may instead affect the motion of the gas once its energy density is comparable with, or greater than, that of the gas. The understanding of the approximate equality of three energy densities is still far from complete and we do not have direct knowledge of whether it holds elsewhere in the disk.

The chapter ended with a brief discussion of star formation. It is suggested that a density enhancement in spiral waves is the main process initiating star formation in our Galaxy and other spiral galaxies at the present time. It is also pointed out that star formation seems more likely to occur as a result of successive fragmentation of a cloud of star cluster mass rather than through the direct formation of individual stars.

7

The chemical evolution of galaxies

Introduction

   A study of the chemical composition of stars in our Galaxy indicates that this composition varies from star to star and that the variation is not random. In particular, there is some correlation between:

   (a) Stellar chemical composition and stellar age, and
   (b) Stellar chemical composition and stellar position or, more accurately, place of origin.

The correlations are in the sense that the oldest stars and the stars formed in the halo region of the Galaxy have a lower heavy element content than the youngest stars and those formed in the disk. It is not completely clear whether this is one correlation or two. To the accuracy to which stellar ages are known *all* halo stars could be older than *all* disk stars and their low heavy element content could just be a consequence of their time of formation. There is, however, also some evidence that halo stars of similar age have very different heavy element content, with those halo stars which were formed furthest from the galactic centre being most deficient in heavy elements. The present estimates of stellar age are not accurate enough to decide whether time of formation is the major factor determining stellar chemical composition or whether there have always been important variations of composition with position.

   As the youngest stars are richer in heavy elements than the oldest stars and as nuclear reactions inside stars converting light elements into heavier elements seem to be the only possible source of the energy which most stars radiate, it seems plausible that the initial chemical composition of the Galaxy was very deficient in heavy elements and that the existing heavy elements were produced in early generations of stars and, having been ejected from stars into the interstellar medium, were then incorporated into new generations of stars. The *hot big bang cosmological theory*, which is in at least good overall agreement with observations of the large scale structure of the Universe, predicts that before any stars or galaxies were formed the chemical composition of the Universe was about three parts hydrogen to one part helium by mass, with only negligible quantities of other elements. The objects in our Galaxy and neighbouring galaxies which have the lowest fraction of heavy elements also appear to have a ratio of hydrogen to helium consistent with the theory. It therefore makes sense to ask whether the present chemical composition of the Galaxy, and other galaxies, can be understood in terms of an initial composition essentially devoid of heavy elements and in terms of changes brought about by nuclear reactions in stars followed by mass loss from stars during galactic history. In this book I shall concentrate on three main questions:

(i) How does the total gas content of a galaxy change with time?
(ii) How does the average chemical composition of the interstellar gas change with time?
(iii) Are there important spatial variations in chemical composition of the gas at a given time?

Most of the discussion will be concerned with our Galaxy because we have the most detailed information about it but I shall also make some remarks about other types of galaxy. The variation of chemical composition from star to star gives important information concerning the second and third questions, as does the present composition of the interstellar gas in galaxies. I start by considering some basic principles. It is believed that stars have been forming from the interstellar gas throughout the history of our Galaxy. At the same time existing stars have been losing matter to the interstellar medium either catastrophically or quietly. Because the mass lost by any star must be less than or equal to its own mass, and in many cases it is significantly less, the net effect of any generation of stars formed is to lead to a reduction in the mass of the *interstellar* medium. Only an infall of matter from the *intergalactic* medium, leading to an increase in the total mass of the Galaxy, can counteract this. Ultimately, the mass lost from the interstellar medium is likely to be *locked* up in dense stellar remnants of three types, black dwarfs, neutron stars and black holes. Although this reduction in mass of the interstellar medium must occur, it is not true that the mass *must* decline monotonically even if no mass is being added from outside. It is easy to understand how this is possible if I consider an extreme example. If, initially, all of the mass of a galaxy were converted into stars, there would then be no interstellar medium but this might subsequently be partially re-established by mass loss from this first generation of stars. In the same way, even if the net effect of the evolution of a single generation of stars is to increase the proportion of heavy elements in the interstellar medium, the fractional heavy element content of the interstellar medium may not increase monotonically. I shall justify this remark on page 166.

Basic principles affecting the chemical evolution of galaxies

A discussion of the evolution of the *gas content* and *chemical composition* of galaxies really involves several distinct factors. They may include:

(i) What causes stars to form? If stars do form what is the distribution of stellar masses?

(ii) What fraction of the mass of a generation of stars is returned to the interstellar medium and when is it returned? What is the chemical composition of this returned mass?

(iii) Is the matter lost by stars mixed thoroughly with the existing interstellar gas?

I comment briefly on each of these questions in turn.

(i) The distribution of stellar masses is usually expressed in terms of an *initial mass function, f(M)*, which is such that the number of stars formed with masses between M and $M + \delta M$ is proportional to

$$f(M)\delta M. \tag{7.1}$$

For young stars in the solar neighbourhood it appears that $f(M)$ is approximately of the *Salpeter form*

$$f(M) \propto M^{-7/3}, \tag{7.2}$$

although there are important deviations from a single power law particularly at high masses. It is not clear what determines this form and whether it can be expected to be more widely applicable. I shall explain later that there are some arguments suggesting that the first generation of stars in our Galaxy may not have had a Salpeter distribution.

In simple discussions of galactic evolution, such as I shall give below, it is usual to suppose that the *rate* of star formation is a function of the density, ρ, of the interstellar gas or alternatively of the surface density or mass per unit area of the disk, σ. It is however clear that, even if stars were to form spontaneously, this must be a gross oversimplification. We should expect the rate of star formation to depend on the temperature, T, and the chemical composition of the gas. In addition it is known that both rotation and magnetic fields may affect the tendency for self-gravitating condensations to form. Although there are indications from observations that stars do form where the gas density is highest (which is hardly surprising) with the rate of star formation depending on something between the first and second power of the gas density, it is clear that this must be a very crude description of what actually happens. In fact, as I have discussed in the previous chapter, there is a strong indication that stars do not tend to form spontaneously where ρ happens to be high but that special factors are necessary to bring the gas to the point where it will collapse and fragment. In particular I have stressed that the compression of gas in spiral density waves probably plays a key role.

(ii) In order to discuss the rate at which gas is returned to the interstellar medium by mass loss from stars, we need a description of the evolution of stars of different masses including their total mass loss in their life history and also some indication of when it occurs. The latter is crucial if we are to have a reasonable idea of the chemical composition of the expelled gas as well as its mass. Mass loss from stars is now known to be common at almost all stages of stellar evolution so that the final mass, particularly of massive stars, may be very much less than the initial mass. The degree of nuclear evolution which occurs in a star depends on this mass loss, as does the composition of material returned to the interstellar medium. It is possible to make estimates of all these quantities and these are used in current discussions of galactic evolution, but they must be regarded as uncertain. In time, additional observations of mass loss from stars and further calculations of stellar evolution should improve the situation.

(iii) The gas lost from stars is returned to the interstellar medium and can be used in the formation of future generations of stars. Does this gas mix smoothly into the existing gas so that we can regard the resulting gas mixture as chemically homogeneous at least locally, or are there significant differences in the chemical composition of the gas within quite small regions of the Galaxy? Furthermore, if the rate of star formation does depend at least approximately on a power of the gas density and if the gas density is not uniform throughout a galaxy, can we expect there to be important differences in chemical composition between widely separated regions of interstellar gas in any one galaxy? Such differences might arise because the greater rate of star formation, in a region where the gas is initially more dense, leads to a more rapid reduction of gas content and increase in heavy element content than in a region where the density is initially lower. Here there are observations of both our Galaxy and other galaxies which may help to check our theoretical ideas and I shall mention them later in the chapter.

A simple model of galactic evolution in the solar neighbourhood

To get a feeling for what this subject involves, I now discuss a highly simplified model of the evolution of the gas content in the disk of our own Galaxy in the neighbourhood of the Sun. I suppose that the gas *in our locality* is, and always has been, fairly well mixed without implying that the chemical composition of the gas is the same throughout the entire Galaxy. Furthermore, I suppose that the really important processes of mass loss from stars concern relatively massive stars, which evolve in a time which is very short compared with any timescale important in galactic evolution. This implies that I assume that the gas ejected by any one generation of stars is available almost instantaneously to be used in a further generation of stars; this is called the *instantaneous recycling* approximation. I also suppose that the rate of star formation does, in fact, depend simply on a power of the local gas density and that the fractional amount of gas returned to the interstellar medium and its *fractional* content of *new* heavy elements is the same for each generation of stars. Implicit in these assumptions is one that the initial mass function is the same throughout the evolution of the Galaxy, but I shall find that there is no need to specify the initial mass function in the equations which I solve. There is one other key assumption of the model which is that the disk formed very rapidly and that no star formation occurred until it had reached its final mass.

This is a very simple model and after I have discussed it I shall wish to ask whether or not it is capable of explaining the principal observational results.

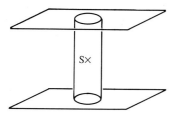

Figure 82.   A cylindrical section of the Galaxy centred on the Sun, **x**.

There are perhaps three key observations:

(a) the variation of chemical composition with stellar age, which gives information about the way in which the chemical composition of the gas in the solar neighbourhood has changed with time,

(b) the numbers of stars of a given type with each chemical composition which, using (a), tells us how the rate of star formation has varied with time and then using the assumptions of the model indicates how the gas density has changed with time, and

(c) independent information about the rate of heavy element enrichment of solar system material derived from studies of heavy radioactive elements. I shall not discuss this observation below.

I shall discuss the results of the observations after I have described the model in more detail.

Now let us look at the model more closely. Suppose that the mass of gas per unit area of the galactic disk is $\sigma(t)$, where t is the time that has elapsed since the disk was formed. In addition let the total mass of stars formed up to time t in a region of the disk with unit cross section (fig. 82) be $\Sigma(t)$. If there were no mass loss from stars $\Sigma(t)$ would necessarily be less than the initial mass of gas in the region $\sigma(0)$. Because mass loss from stars allows material to be used in more than one generation of stars, $\Sigma(t)$ need not be less than $\sigma(0)$. It is convenient to define two new quantities $\mu(t)$ and $S(t)$ by the equations

$$\sigma(t) = \sigma(0)\mu(t) \tag{7.3}$$

and

$$\Sigma(t) = \sigma(0)S(t), \tag{7.4}$$

so that $\mu(t)$ and $S(t)$ are dimensionless measures of the gas density and total mass of stars formed, respectively, with $\mu(0) = 1$ and $S(0) = 0$. If the rate of star formation depends on some power of the surface density of the gas,† I have

$$dS/dt = C\mu^n, \tag{7.5}$$

† I have previously assumed that the rate of star formation depends on a power of the volume density ρ not the surface density. The two assumptions are equivalent if the thickness of the disk does not change with time.

where C and n are constants. As has been mentioned previously there is some indication from observations that n might satisfy $1 \leqslant n \leqslant 2$.

As a result of successive generations of stars, the fractional heavy element content (by mass) of the interstellar gas, which I write $Z(t)$, gradually changes from its assumed initial value $Z(0) = 0$. In fact the material in the disk is probably slightly enriched by mass loss from halo stars which formed before the disk. As the heavy element content of halo stars is very much less than that of the average disk star and as the visible mass of the halo is very much less than that of the disk, there should not be much error in ignoring this enrichment. I shall assume that in each generation of stars a fraction α of the mass is not returned to the interstellar medium; thus this fraction is contained in the remnants of exploded stars or in the low mass stars whose evolution time is longer than the age of the Galaxy. The fraction $1 - \alpha$ is assumed to be returned to the gas instantaneously. I assume further that a fraction λ of the mass of each generation of stars is contained in regions in stars which are both completely converted into heavy elements *and* ejected into the interstellar medium. My basic assumption is that α and λ as well as C and n are constants and I now have enough information to determine a relationship between the mass of gas remaining at any time in galactic evolution and its chemical composition, as I shall now demonstrate. Note that in representing the heavy element content by a single parameter Z I am implicitly assuming that all stars have the same mix of heavy elements. Although this is not true in detail, it is a reasonable first approximation.

Initially all of the matter is gas. Of each generation of stars a fraction α is not returned to the gas and what is returned is assumed to be returned instantaneously. Thus at any time

$$\mu = 1 - \alpha S. \tag{7.6}$$

Combining (7.6) with (7.5), I have

$$d\mu/dt = -\alpha dS/dt = -\alpha C\mu^n = -\mu^n/t_0, \tag{7.7}$$

where $t_0 \equiv 1/\alpha C$ has the dimensions of time and can be regarded as a characteristic time for significant change in gas density. Now consider the enrichment of the interstellar medium with heavy elements. The total quantity of heavy elements changes for two reasons. Star formation uses up heavy elements as well as hydrogen and helium and heavy elements are contained in the mass which is lost by stars. In the latter case some of the heavy elements were present when the stars were formed and some were produced in the stars themselves. The total amount of heavy elements in unit mass of the disk at any time is $Z\mu$ and we have

$$\frac{d(Z\mu)}{dt} = -Z\frac{dS}{dt} + (1 - \alpha - \lambda)Z\frac{dS}{dt} + \lambda\frac{dS}{dt}. \tag{7.8}$$

The three terms on the right hand side of equation (7.8) arise as follows. The first term is the rate of loss of heavy elements due to star formation. The second term represents the return to the interstellar medium of matter whose heavy element content has not been changed inside stars; the instantaneous recycling approximation implies that this matter has the current value of Z rather than a value appropriate to some time in the past. The third term describes the return to the interstellar medium of matter which has been completely processed into heavy elements. Equation (7.8) can equally be written

$$d(Z\mu)/dS = \lambda(1 - Z) - \alpha Z. \tag{7.9}$$

I can now eliminate S by using equation (7.6). If I do this I have

$$\frac{d(Z\mu)}{dS} = \mu\frac{dZ}{dS} + Z\frac{d\mu}{dS} = -\alpha\mu\frac{dZ}{d\mu} - \alpha Z,$$

so that

$$-\mu dZ/d\mu \equiv dZ/d\log_e(1/\mu) = \lambda(1 - Z)/\alpha. \tag{7.10}$$

λ/α is usually called the yield of heavy elements; it is the ratio of the mass completely converted into heavy elements to the mass locked up in stars. If I write $\lambda/\alpha = p$ and note that at all times up to the present in the life history of the Galaxy $Z \ll 1$, (7.10) can be written to sufficient accuracy

$$dZ/d\log_e(1/\mu) = p$$

which integrates into

$$Z = p\log_e(1/\mu), \tag{7.11}$$

which, for the simple model, is a relation between the heavy element content of the interstellar gas and the mass of gas remaining and which is independent of n. It is clear from equation (7.10) that the assumption $Z \ll 1$ is not needed to obtain an n-independent result. If I wish to see how Z and μ vary with t, a value for n is required. Equation (7.7) can be integrated to give μ as a function of t:

$$\left.\begin{array}{l} \mu = \exp(-t/t_0), \quad n = 1, \\ \mu^{n-1} = 1/[1 + (n - 1)t/t_0], \quad n > 1. \end{array}\right\} \tag{7.12}$$

If I now call the present time $t = t_1$, so that at present $Z = Z_1$ and $\mu = \mu_1$, I can use the expressions which I have obtained to calculate some quantities which can be compared with observation. Consider in particular long-lived stars which are of such a low mass that even those formed in the earliest phases of galactic history are still main sequence stars today. This is certainly true for all stars slightly less massive than the Sun. I can use our model to predict what should be the relative numbers of these stars with all values of Z between 0 and Z_1. Rather more crudely I can estimate what should be their average heavy element abundance.

The total number of stars formed with heavy element abundances $\leq Z$ is proportional to $S(Z)$, the total mass of stars formed up to the time when the heavy element abundance is Z. The total number formed up to the present time is similarly proportional to S_1. Now from equations (7.6) and (7.11),

$$\frac{S(Z)}{S_1} = \frac{1 - \mu}{1 - \mu_1} = \frac{1 - \exp(-Z/p)}{1 - \exp(-Z_1/p)} = \frac{1 - \mu_1^{Z/Z_1}}{1 - \mu_1}. \tag{7.13}$$

Note that this result does not depend on p. Thus if I can observe μ_1 and Z_1 today, I can predict the distribution of long-lived stars of different heavy element abundance according to this model and I can compare this with observations. The rather cruder property is the average heavy element abundance of all low mass stars formed up to the present time. This is

$$\langle Z \rangle_1 = \int_0^{S_1} Z(S) dS/S_1$$

$$= \int_0^{\mu_1} Z(\mu) d\mu/(1 - \mu_1)$$

$$= p[1 - \{\mu_1/(1 - \mu_1)\} \log_e (1/\mu_1)]. \tag{7.14}$$

Once again I can eliminate the yield, p, by using equation (7.11) at the present time to obtain

$$\frac{\langle Z \rangle_1}{Z_1} = \left[\frac{1}{\log_e (1/\mu_1)} - \frac{\mu_1}{1 - \mu_1} \right]. \tag{7.15}$$

Comparison with observations

This completes my description of what can be regarded as the simplest possible model of chemical evolution of our neighbourhood of the Galaxy. I must now compare it with observation. It is not possible to make a complete comparison with observation without possessing values for n, α and p but as I have just shown equations (7.13) and (7.15) contain no quantities which cannot in principle be obtained directly from observation and this fact can be used to give a test which is independent of n, α and p. To use it I need values of μ_1 and Z_1 and information about the numbers of stars with different heavy element abundances. I have already explained in previous chapters that the exact division of the mass in the solar neighbourhood between gas and stars is rather uncertain. From the discussion of *Oort's limit* given in Chapter 6 we know that there is missing matter in the solar neighbourhood which may conceivably be at least partially in the form of gas. What seems quite clear is that $\mu_1 > 0.1$ and that $\mu_1 \approx 0.2$ is quite possible. It will appear shortly that no more accurate value of μ_1 is required to show that the simple model is not in good agreement with the observations. The value of Z_1 must be obtained from the interstellar gas and/or recently formed stars. Here there is some significant scatter in the results and an average value must be used for Z_1; further comments on this scatter will be made below.

It is not possible to describe the observations of chemical compositions of stars in detail in this book. Fig. 83 illustrates the results obtained and their comparison with the theoretical predictions of equation (7.13). The solid lines show the relation between S and Z for two values of μ_1 (0.1, 0.2), while the solid circles are observational points. It is clear that the general character of the theoretical curves is totally different from that of the observations. In particular the observations

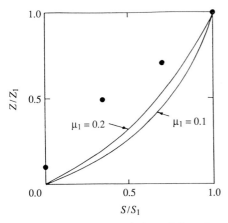

Figure 83. Chemical evolution of the solar neighbourhood. The heavy element content in terms of its current value is plotted against the mass of star formed in terms of its current value. The filled circles are observations. The curves show predictions of the simple model for two values of the current gas fraction α_1.

give far too few stars with really low heavy element abundance, in comparison with the theoretical curves.

I must now consider what modifications are required to our discussion in order to bring theory and observation into agreement. There are two basic ways in which this can be done. One involves relaxing some of the simple assumptions of our model, which will inevitably have the effect of making it more complicated, and the other involves casting doubt on the relevance of the observations. Let us consider the latter point first.

The stars which are plotted in fig. 83 are low mass stars which are situated close to the Sun. In making a comparison with the theoretical model, I am assuming that the stars that are observed are a fair sample of the stars in that mass range which have been formed in the solar neighbourhood during the whole history of the galactic disk. In Chapter 2 I have explained that old stars occupy a thicker disk than young stars. This means that, to obtain an estimate of the total number of stars formed per unit area of the disk, a larger volume must be studied for old stars than for young stars. Some astronomers believe that an inadequate correction for this effect has been made in the observations shown in fig. 83 and that there is not a serious deficit of highly metal-deficient low mass stars. Other astronomers are convinced that the problem remains and that our model must be modified in some way.

At present I have only used the cumulative numbers of stars with heavy element abundance less than a certain value and I have not asked whether the stars with low heavy element abundance really are the oldest, as they must be according to the simple model. It is not easy to obtain accurate values of stellar ages but, to the accuracy with which these can be obtained, there does appear to be a slow

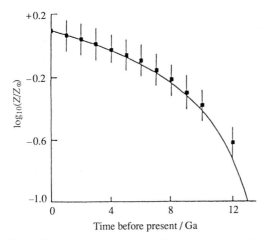

Figure 84.   A schematic plot of heavy element abundance versus age for individual stars. (1 Ga is 10^9 yr.)

monotonic increase of heavy element abundance with time of formation during galactic history, following a rapid initial rise. Superimposed on this average trend there are very large fluctuations, as shown schematically in fig. 84. The scatter of Z at given t might be reduced if it were assumed that all stars with too high Z were younger than they appear by the typical uncertainty in stellar ages and conversely for stars with too low Z, but this could not remove the scatter in the chemical composition of very young stars and the interstellar medium today, which has already been mentioned. This scatter can be explained if our assumption that the interstellar medium is always well mixed is wrong. There is also the possibility that our assumption that the surface chemical composition of a star is always the same as the interstellar gas out of which it was formed might be seriously wrong. If this were true it would invalidate our comparison of observations with our simple model and it might also account for the absence of many disk stars with low heavy element abundance, as I shall explain shortly.

When we study the surface chemical composition of a star we are studying one of three things:

(i)   the chemical composition of the star when it was born,
(ii) the original composition as modified either by nuclear reactions inside the star or by some segregation of elements produced by motions inside the star,
(iii) the composition of interstellar matter recently accreted by the star.

In order to eliminate modifications of composition due to nuclear reactions, we study main sequence stars in which no significant nuclear reactions have occurred. Internal segregation of elements almost certainly occurs in only a very small minority of stars, and (iii) has generally been supposed to be unimportant for most stars. It cannot be overwhelmingly important otherwise *all* stars would have a surface chemical composition similar to that of the interstellar medium in their neighbourhood and no disk stars would be observed to have a low heavy element abundance. Indeed, it is known both that accretion from the general interstellar medium is ineffective and that many stars, including the Sun, are continuously losing mass at a very low rate, which has the effect of keeping the surface of the star clean.

The belief that (iii) is always unimportant has been questioned because of the discovery that galactic spiral arms contain gas clouds which are much denser than the average interstellar medium. As the spiral arms are a wave pattern which sweeps through the distribution of stars, all stars will periodically pass through spiral arms and on some of these passages they may pass through very dense clouds. In the case of the Sun, we can expect it to pass through a spiral arm about once every 10^8 years. It has been suggested that accretion of gas from these dense clouds might be much more effective than from the general interstellar medium and, if a star passed through such a cloud, its surface chemical composition would no longer give information about its composition at birth. The effect would be greatest for old stars with low initial heavy element abundance which have passed through spiral arms most times and the absence of a large number of stars with a low observed heavy element content could then perhaps be understood.

Although this suggestion is in principle quite attractive, detailed discussions suggest that it is unlikely to be effective enough to explain the observations. If stars are normally losing mass, this mass loss must be reversed and replaced by accretion when a star passes through a dense cloud and it is not clear that this will happen. Even if accretion does occur, subsequent mass loss must not remove the accreted material rapidly. In addition, if the total mass accreted is small, in low mass stars convection in the atmosphere will mix it with the original stellar material and this will reduce the effect on the observed composition. My own feeling is that accretion is unlikely to be important and that the surface chemical composition of old stars does tell us something about their origin.

Modifications to the simple theory

I now turn to ways in which the simple model can be modified to try to bring it into agreement with the observations. Clearly one assumption which can be questioned immediately is that all generations of stars have similar properties. Star formation is as yet imperfectly understod but it is certainly an oversimplification to say that the rate of star formation depends only on the gas density and that the mass distribution of stars formed has a unique form. I have already said on page 149 that both the rate of star formation and the initial mass function must be influenced by the temperature (or random kinetic energy) and chemical composition of the gas as well as its angular momentum and magnetic field. In addition, the observation that young stars are preferentially found in and near spiral arms suggests that star formation is largely triggered by the passage of a spiral wave rather than being something which occurs spontaneously. However, the final average rate of star formation (over a period of galactic rotation, for example) and the mass function *might* be insensitive to some of these factors and I shall not attempt to discuss the problem in detail.

Attempts to reconcile theory and observation have taken into account three factors which might be important:

(i) Prompt initial enrichment,
(ii) Metal enhanced star formation,
(iii) Infall.

I shall now define and discuss each of these in turn. The first two will be seen to take account in some approximation of some of the factors mentioned in the last paragraph.

Prompt initial enrichment

If our general ideas are correct, when the Galaxy was first formed it contained essentially no heavy elements, but we do not at present observe disk stars with this essentially zero heavy element content. This could be explained because all disk stars have a contaminated outer layer as suggested on page 157, but there at least three other possible explanations. Two of these involve a consideration of the chemical composition of objects in the galactic halo, such as the globular star clusters. Stars in globular clusters are found to have very much lower Z than most disk stars but the value of Z varies from cluster to cluster. In addition those clusters which appear to have been formed nearest to the centre of the Galaxy tend to have higher Z than those formed far out. Finally, no stars have been observed which are devoid of heavy elements. In the simplest theories of galaxy formation it is supposed that a pregalactic gas cloud contracted under gravity and that during this initial collapse subcondensations formed which became the globular clusters. Because the pregalactic cloud was rotating it did not collapse to a point but flattened to form a disk. If no energy were dissipated in this collapse the cloud would have been able to re-expand to its original radius. In fact we expect the uncondensed gas to dissipate a large amount of energy through collisions when the system is very thin and afterwards to form a thin disk. In contrast the protoclusters being much denser than the remainder of the gas could pass through the disk without losing much energy and would continue to occupy a much larger volume as is observed.

Although this general picture is very attractive it does not explain how the globular clusters acquired their heavy elements and why those nearer to the centre of the Galaxy have the larger share. One possible explanation is related to the idea that the Galaxy is much more massive than is generally believed and that star formation could initially have occurred in a massive halo, as has already been mentioned on page 125. Mass loss from these extreme halo stars which might have had a very different mass function from present day disk stars could then add processed matter to the remainder of the galaxy before either the globular clusters had completed their contraction and started to form stars or the first generation of disk stars formed, so that this might account for the absence of disk stars with very low heavy element abundances and explain why globular cluster stars have any heavy elements at all. The second suggestion is that the globular clusters did not form during the initial contraction of the protogalaxy. Instead it is suggested that much of the Galaxy collapsed in gaseous form to form a central nuclear bulge and that heavy elements produced in stars or more massive objects enriched the remainder of the halo and the disk. This would explain both the presence of heavy elements when the first stars in the solar neighbourhood or the globular clusters were formed and the occurrence of higher heavy element abundances near to the centre of the Galaxy. There is a spread in globular cluster ages and there is therefore also a possibility that the younger globular clusters contain heavy elements which were produced in stars in the older clusters.

The third possible explanation of the lack of *disk* stars with very low Z is that the first generation of disk stars (in a Galaxy without significant previous star formation in a massive halo) was essentially all massive stars. If this were so, the first generation of massive stars would produce heavy elements but leave behind no low mass, low Z stars which could be observed today. On this model, much of the mass of the disk must at present be in the form of black hole remnants. Although this suggestion sounds totally *ad hoc*, there is some theoretical support for the idea that star formation in a cloud of pure hydrogen and helium leads to a very different initial mass function from star formation in a cloud containing heavy elements, with more massive stars being preferred.

All three of these suggestions lead to non-zero Z when the first disk stars are formed and this process is called *prompt initial enrichment*. In calculations of galactic evolution the degree of enrichment tends to be taken as a free parameter which is not necessarily immediately related to one of the mechanisms which I have mentioned. Although I shall not discuss the topic in this book, a full discussion of galactic chemical evolution must eventually give a self-consistent model of disk, nuclear bulge and halo.

Metal-enhanced star formation

Cool gas clouds are more likely to collapse to form stars than hotter gas clouds because it is the random kinetic energy of clouds that resists the gravitational attraction. In addition, the presence of heavy elements in even a small amount enhances the cooling rate of interstellar gas. At any time the interstellar medium does not have a completely homogeneous chemical composition; we know this from direct observations today and it is strongly suggested by the variations in chemical composition in stars of a given age. Perhaps stars are formed preferentially in regions of higher than average heavy element abundance, so that we always expect the average Z of stars of a given age to be higher than the average Z of the interstellar medium out of which they were formed. This process is called metal-enhanced star formation. This idea goes some way towards explaining the low numbers of stars with low Z but it cannot solve it completely if the enhancement is related to the average abundance. When Z is actually zero no fluctuations are possible and when Z is very small the fluctuations will also be small.

Infall

The third sugestion for a modification of the simple model considers the possibility that the Galaxy is not an isolated system and that it is accreting matter from the *intergalactic medium*. Such matter is assumed to be unprocessed and to contain no heavy elements. This matter could either be genuine intergalactic material through which the Galaxy is passing or it could be matter which has always been, in principle, part of our Galaxy but which, being from the outer region of the protogalaxy, has taken longer to collapse to the disk than has the bulk of the mass. It is easy to obtain a qualitative idea of the effect of *infall* of intergalactic material. If some gas with no heavy elements is added to the interstellar medium, the fractional amount of heavy elements, Z is reduced. If, at the time that infall is occurring, the net effect of star formation and mass loss from stars is to increase Z, infall will either reduce the rate of increase of Z or, if it is very intense, cause Z to decrease. This tendency of infall to reduce Z will be reinforced if infall is sufficiently prolonged that it is still important when mass loss from low mass stars discussed on page 166 is relevant.

One form of infall which may go a long way towards explaining the shortage of stars with low heavy element content in the galactic disk is the following. Suppose that star formation started in the disk at a time when only a small fraction of the final mass of the disk had collapsed to the plane. There would then be only a relatively small number of stars with very low heavy element content because the heavy elements produced in this first generation of stars would be available to be incorporated in further stars formed as the galactic disk increased in mass. An appropriate combination of initial mass function and rate of accretion of matter to the disk might be able to explain the numbers of stars with different values of Z.

Solar neighbourhood infall

The general belief today is that star formation did start in the galactic disk before it had reached its final mass and, in fact, that infall may have been very important throughout past galactic history. I now discuss a simple model of the chemical evolution of the solar neighbourhood which incorporates infall. Because I now assume that the disk starts with zero mass, I can no longer normalise the surface gas density σ and the star density Σ with respect to the (constant) total density of the disk. Instead I normalise quantities with respect to the present surface gas density σ_1. I also choose to work in terms of the total mass locked up in stars $\alpha\Sigma$ rather than Σ itself. Thus I write

$$\sigma = \sigma_1 x, \tag{7.16}$$

$$\alpha\Sigma = \sigma_1 y. \tag{7.17}$$

The star formation law can then be written

$$\frac{dy}{dt} = \alpha C'x^n, \tag{7.18}$$

where C' is a constant related to C in equation (7.5).

If mass infalls to the disk at a rate a per unit area per unit time, the rate of change of surface gas density can be written

$$\frac{dx}{dt} = -\alpha C'x^n + (a/\sigma_1), \tag{7.19}$$

where the first term on the right hand side in (7.19) represents the loss of gas due to star formation and the second term the gain due to infall. In addition the equation corresponding to (7.8) now takes the form

$$\frac{d(Zx)}{dt} = -Z\frac{dy}{dt} + p(1 - Z)\frac{dy}{dt}. \tag{7.20}$$

There is no infall term in this equation because it is assumed that the infalling matter has no heavy elements. If a new time variable $\tau = \alpha C't$ is introduced, equations (7.18) to (7.20) can be written

$$\frac{dy}{d\tau} = x^n, \tag{7.21}$$

$$\frac{dx}{d\tau} = -x^n + b, \tag{7.22}$$

$$\frac{d(Zx)}{d\tau} = (p(1 - Z) - Z)\frac{dy}{d\tau} = (p(1 - Z) - Z)x^n, \tag{7.23}$$

where $b = a/\alpha C'\sigma_1$.

There is no simple solution to these equations in the case of a general value of n or infall function b. The solution can be approached by assuming that the observations of low mass stars give a relation between Z and y of the type shown in fig. 83. If such a relation is assumed known, equation (7.23) can be integrated to give

$$x = (1/Z)\int_0^y [p(1 - Z) - Z]\, dy'. \tag{7.24}$$

Once x is known as a function of y from equation (7.24), equations (7.21) and (7.22) combine to give

$$b = x^n + \frac{dx}{d\tau} = x^n\left[1 + \frac{dx}{dy}\right]. \tag{7.25}$$

If a value of n is then assumed for the star formation law, equation (7.25) then gives b as a function of y. There remains one point which I have not fully explained. Fig. 83 gives Z/Z_1 as a function of S/S_1 but I have normalised the mass

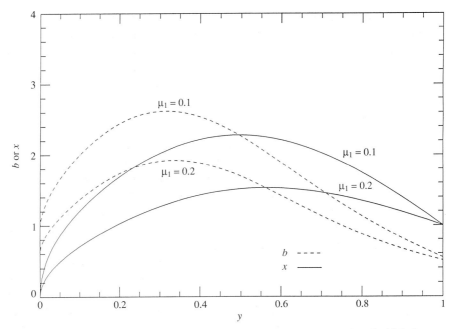

Figure 85.   A model for the chemical evolution of the solar neighbourhood which fits the observed data shown in fig. 83. The evolution of the gas infall function b and the gas surface density x are shown.

locked up in stars with respect to the current gas density in the disk. The current value of y is related to the current gas fraction of the disk so that

$$y_1 = (1 - \mu_1)/\mu_1. \tag{7.26}$$

Calculations have been performed for a particular catalogue of metallicities of low mass stars, for $n = 1$ and for $\mu_1 = 0.2$ and the resulting plots of x and b as a function of y are shown in figure 85. These results (and similar ones with $\mu_1 = 0.1$ and $n = 2$) suggest that the observed distribution of heavy element abundances of low mass stars can be best understood if infall has been important throughout past galactic history and that it continues to be important today. These results would of course be modified if the deficiency of low mass stars with low heavy element abundance proves to be more apparent than real. It does however indicate the manner in which infall of gas can modify the predictions of the simple model. It should be stressed that, if these infall models provide even an approximately correct model for the solar neighbourhood, the local luminosity and star forma-tion rate has never been orders of magnitude higher than the current rate. Whether or not this is also true for the Galaxy as a whole depends on models for the evolution of complete galaxies to which I now turn.

Present observations are not clear concerning infall of gas to the galactic disk. There are some clouds of gas observed to be moving towards the disk with a high velocity. It is not clear whether these high velocity clouds represent matter which is reaching the disk for the first time or whether it is material which has been thrown out of the disk or the nuclear bulge and is now returning. There are different estimates of the amount of matter in these clouds but the highest would imply that infall could be important at the present time.

Variations of chemical composition in galaxies

There is one other type of observation which might help to differentiate between the various possible explanations of the chemical evolution of this and other galaxies. This is an observation of how the chemical composition varies from point to point within galaxies at the present time, combined with observations of the relative amounts of mass in the forms of gas and stars at different points. Observations of these types are becoming available. It is not possible to make detailed studies of the chemical compositions of stars in galaxies other than our own because the stars are too faint, but a study can be made of gas clouds. Even in our own Galaxy the major progress is in a study of gas clouds, because of the strong absorption of starlight in the galactic plane. What evidence there is at present suggests that there are significant radial composition gradients across galaxies in the sense that gas clouds near the centres of galaxies are richer in at least some heavy elements than those nearer the edge. In addition the variation of composition appears to be rather smooth. If this latter point is confirmed it will be an argument against metal enhanced star formation as there may be no metal-rich clouds in which such a process can occur.

From the theoretical point of view, radial composition gradients should cause no surprise. Thus I could generalise my discussion which has been concerned with the properties of the gas and stars in the solar neighbourhood. Consider first the simple model, even though we know that this cannot explain the properties of the solar neighbourhood. On that model the heavy element abundance is directly related to the mass of gas remaining through equation (7.11), assuming that the yield p is universal. We should therefore expect to see high heavy element abundances where there is less gas; unfortunately the dependence is only logarithmic so that it would not be easy to extract significant results because of observational errors and genuine observational scatter. In making this prediction, I have assumed that the gas in each locality in the Galaxy can be regarded as behaving independently of the gas in any other region. This is no more than the assumption which I have already made in studying the solar neighbourhood alone. I should in fact study the efficiency of mixing of gas between different regions of the galaxy. A simple argument – based on how far mass ejected by supernovae travels before it is mixed into the general interstellar medium and on the observed random velocities of interstellar clouds – suggests that, although some mixing will occur, the assumption of no mixing is likely to be a much better approximation than the assumption of complete mixing. There is however the possibility of radial motions in disks which I shall mention below.

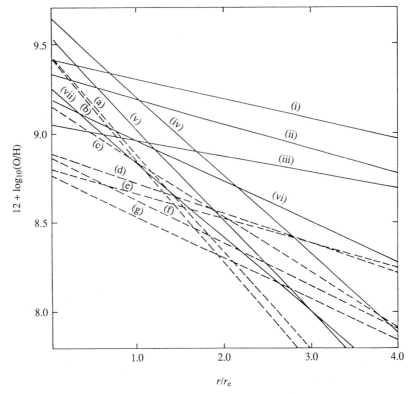

Figure 86.   Radial abundance gradients of oxygen in spiral galaxies. Solid lines are early type spirals and dashed lines are late types. The individual identifications are: (i) NGC4321, (ii) M83, (iii) NGC1365, (iv) M51, (v) Galaxy, (vi) M31, (vii) NGC2997, (a) I342, (b) NGC6946, (c) M101, (d) M33, (e) NGC2403, (f) NGC2793, (g) NGC300. (O/H) is the ratio of number densities of oxygen and hydrogen and r_e is the scale length of the exponential disc.

Radial abundance gradients in galaxies

Some observational data about abundance gradients in galaxies are shown in fig. 86. The gradients are in some cases very steep and it has not proved easy to produce theoretical models whose gradients agree with observations. As I have already explained, although gradients are predicted in the simple model, the heavy element abundance depends only logarithmically on the fractional gas density. In addition infall may act to reduce gradients rather than to increase them. There are various possible ways in which the gradients may be increased. One is to assume that there was some form of prompt initial enrichment,which was concentrated in the central regions of galaxies and which might be associated with the formation and early evolution of a central nuclear bulge which I have mentioned earlier. In that case abundance gradients in disks might be related to their birth conditions.

A second possibility is to assume that the time of formation of the disk is dependent on distance from the centre of the galaxy. In these models, which go by the name of biased disk formation, it is assumed that material with low angular momentum per unit mass falls to the centre of the galaxy much more rapidly than material with higher specific angular momentum falls to the outer regions of the disk. If that is correct, the long timescale for formation of the solar neighbourhood, which I have just discussed, need not be characteristic of regions nearer to the galactic centre and the outer regions of the disk may still be only partially formed today. Calculations of disk evolution with biased infall have given higher abundance gradients, nearer to those observed, than those obtained when infall proceeds at the same rate at all radii.

Radial flows in galaxies

A third possibility is that radial gas motions have been important during disk history. If the gas moves inward, while the stars do not, or if the stars move more slowly than the gas, this can modify and increase abundance gradients. Two different reasons for such radial motions have been suggested. The first is the effect of viscosity in the interstellar gas. The galactic disk is differentially rotating. Collisions between interstellar gas clouds will lead to exchanges of angular momentum. The observed differential rotation is, however, demanded by the mass distribution of the Galaxy, if the Galaxy is to be in a steady state. The mass distribution can itself only be changed by radial flows, although, if most of the mass is in an invisible halo, motions in the disk will not have much effect on the angular momentum required for circular orbit. The effect of cloud collisions will lead to some inward radial motion of the gas which will not be shared by the stars because their collisions are so rare.

Radial motions can also arise as a result of the manner in which a galaxy forms. It is possible that when matter first reaches the disk it does not possess the correct angular momentum for circular orbit at the radius at which it finds itself. If it has too little angular momentum it will move inward; if too much it will move outward. If material is in centrifugal equilibrium at some moment, but the mass of the galaxy subsequently increases, it will cease to have the appropriate angular momentum and will move radially inwards until equilibrium is again attained. It is possible that the early formation phase of galaxies involves steady radial flows as matter tries to attain the centrifugal equilibrium which is observed in galaxies today. Calculations of the chemical evolution of galaxies with radial flows have given increased abundance gradients compared with models with similar accretion rates but with no radial flows.

All of this discussion of the radial abundance gradients in galaxies is closely bound up with models for the formation and early evolution of galaxies. Unfortunately, direct observations of galaxy formation are not easy because there is as yet no clear evidence that there has been more than one epoch of significant formation of galaxies. Forming galaxies can probably only be observed at high redshift when only their integrated properties are available. Ultimately a discussion of the chemical evolution of a galaxy such as our own must give a unified treatment of the halo, disk and central nuclear bulge.

Non-instantaneous recycling

All our detailed discussion has made use of the instantaneous recycling approximation, which says that all of the important exchange of gas between stars and the interstellar medium occurs from relatively massive stars which evolve very rapidly compared with the general timescale of galactic evolution. Clearly this is at best an approximation as some mass loss does occur from low mass stars which do not themselves produce much in the way of heavy elements. I have also assumed that all of the gas which is expelled from stars is retained in the galaxy. Some remarks will now be made about each of these assumptions in turn.

Mass loss *does* occur from low mass stars at late stages in their evolution. It is observed for example that *planetary nebulae* (fig. 87) are formed by mass loss from stars of around a solar mass and upwards and there are also arguments from a study of observed properties of stars in globular star clusters combined with theories of stellar evolution which suggest (but not completely conclusively) that all stars in this mass range may eject more than 20 per cent of their mass at a relatively late stage in their evolution. If the distribution of stellar masses is anything like the Salpeter mass function (7.2), most of the mass of any generation of stars is in the form of low mass stars. This suggests that the gas lost by low mass stars is an important addition to the interstellar gas, which is certainly not adequately covered by the instantaneous recycling approximation. This is particularly true when the galaxy is about 10^{10} years old or older because by then the low mass stars are beginning to complete their evolution and the total mass of remaining interstellar gas is low. Because the fractional mass of gas is lowest in the central regions of galaxies, the effect of non-instantaneous recycling first becomes important there. This is another process which tends to flatten abundance gradients, because of the effect which I mention in the next paragraph.

The mass lost by low mass stars *can* be included in a model of galactic evolution as an effective infall term except that it is not now infall of completely unprocessed material as the stars were not formed out of unprocessed material. In addition, although really heavy elements are not formed in these low mass stars, they may make an important contribution to the production of ^{12}C and ^{16}O. Earlier in the chapter it was mentioned that the fractional heavy element content of the interstellar medium need not increase monotonically with time even in the absence of infall. It is easy to see how mass loss from low mass stars *might* lead to a situation in which $Z(t)$ initially increases and later falls. If most of the mass of a galaxy is first converted into a single generation of stars, the initial effect of mass loss from massive stars will be to produce an interstellar medium with a relatively high Z. If this is later followed by the loss of (essentially) unprocessed material from low mass stars, Z will fall. Thus infall of intergalactic matter is not needed to destroy a monotonic increase of $Z(t)$. An observation of a monotonic increase of Z serves as a constraint on both the amount of infall and the rate of consumption of the galactic gas.

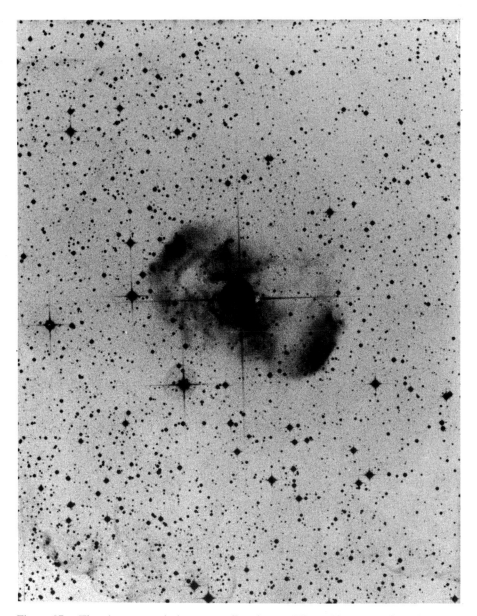

Figure 87.   The planetary nebula surrounding the star HD148937, a negative print.
(Photograph taken on the UK 1.2 m Schmidt Telescope reproduced by permission of the
Royal Observatory, Edinburgh.)

Elemental and isotopic abundances

The study of interstellar molecules which has been mentioned on page 49 is
begining to give very useful information about the *isotopic abundances* of the light
elements. Thus isotopes such as deuterium, ^{13}C, ^{15}N, ^{17}O and ^{18}O can be detected
in some molecular clouds. Relative abundances such as $^{13}C/^{12}C$ and $^{15}N/^{14}N$

apparently vary from place to place in the Galaxy. This is not unexpected because, for example, the relative production of the isotopes is different in stars of different masses. At varying distances from the galactic centre the mix of stars which have produced the observed heavy elements will not have been the same. Thus we can expect to observe variations in $^{13}C/^{12}C$ and $^{15}N/^{14}N$. We must now require our models to predict the observed values of these ratios. Some models of galactic evolution which satisfy the tests which we have previously mentioned do not produce the correct isotopic abundances and can therefore be discarded.

In this book I have treated the chemical elements other than hydrogen and helium as a single entity. I have simply referred to the heavy element abundance Z. This is not of course an adequate description of what actually happens in galaxies. Stars of different mass produce and expel into the interstellar medium a different mix of heavy elements. Some elements which are called *primary* can be produced directly from nuclear reactions involving only hydrogen and helium, whereas other elements called *secondary* can only be produced from other elements which have themselves first been formed from hydrogen and helium. It is observed, for example, that the abundance of nitrogen relative to oxygen varies both from point to point in galaxies today and between young stars and old stars in our Galaxy. Ultimately an understanding of the chemical evolution of galaxies must involve an explanation of the properties of the abundances of all of the individual elements.

Gas retention by galaxies

Consider next the question of gas retention by a system of stars such as a star cluster or a galaxy. If a system of stars is of sufficiently low mass, it scarcely seems possible for it to have more than one generation of stars. It retains no gas and has no chemical evolution. Galactic (open) star clusters in our Galaxy certainly fall into this category. Whether or not all of the gas of the protocluster is turned into stars in the initial burst of star formation, it will be expelled from the cluster either by the heating effect of ultraviolet radiation from massive stars or by mass ejected with high velocities by a single supernova. Subsequently, as the escape velocity from any star is greater than the escape velocity from such a cluster, most mass lost by stars will escape freely from the cluster. Certainly there is no evidence of significant variations of chemical composition indicating several generations of stars in galactic clusters.

When a stellar system is sufficiently massive it becomes possible for it to retain gas. Observations of chemical compositions of stars in nearby systems suggest that the critical mass is possibly about $10^6 M_\odot$ or a little more, so that the most massive globular clusters and even dwarf galaxies exceed this value. Whether a galaxy then retains gas and has many generations of stars appears to depend critically on how efficiently gas is turned into the first generation of stars. If most of the gas is not immediately turned into stars, the remaining gas will be effective in retaining the gas expelled by the first generation of stars and the galaxy may then retain a significant amount of gas for much of its lifetime. This appears to be the case in our own Galaxy and other spiral galaxies and in irregular galaxies such as the Small and Large Magellanic Clouds. The same does not appear to be generally true for elliptical galaxies. As I have already explained in Chapter 3 they have little or no gas and few young stars. It appears that the reason for this is that star formation was extremely efficient when elliptical galaxies first formed and they have not subsequently been able to retain substantial amounts of gas. Three comments should however be made on this apparent difference between elliptical and spiral galaxies.

The first refers again to the massive haloes to spiral galaxies which have been mentioned frequently in this book. If these are composed of ordinary matter rather than of weakly interacting particles, star formation in elliptical galaxies may not have been quite so much more efficient than initial star formation in spiral galaxies as at first appears. The second comment is that some elliptical galaxies have bright blue cores indicating the presence of massive main sequence stars whose main output of radiation is in the blue and ultraviolet region of the spectrum. This shows that gas can *sometimes* accumulate in the central regions of elliptical galaxies and produce new massive stars. The third point is that, as I have already mentioned in Chapter 3, some elliptical galaxies are associated with intense radio sources and the energy powering the radio source is believed to come from the centre of the galaxy. The gravitational energy released by infall of gas into the galactic nucleus is one strong candidate for the source of this energy. Once again, this suggests that some gas does accumulate in the centres of elliptical galaxies and that it may not necessarily fragment to form stars.

Removal of gas by the intergalactic medium

I have earlier suggested that galaxies may accrete intergalactic matter but there is a further possibility. If a galaxy is moving sufficiently rapidly through a (relatively) dense intergalactic medium the intergalactic gas might push the galactic gas out of the galaxy instead of being accreted. This is likely to be most important in clusters of galaxies. In several cases X-ray emission from clusters of galaxies has been interpreted as being due to hot intergalactic gas with a high enough density to be able to sweep some galaxies clean. Although this may not be a particularly important factor in the difference between elliptical and spiral galaxies, it may have some bearing on the differences between S0 and ordinary spiral galaxies.

Differing rates of evolution of galaxies

One further observation which must eventually be explained is the differing rate of evolution of galaxies of different types which do retain their gas. The Magellanic Clouds for example have a larger fraction of their mass in the form of gas than does our Galaxy and their fractional content of heavy elements is lower. They thus have the appearance of being younger systems although their actual ages are believed to be similar to that of the galaxy. Although the

Magellanic Clouds must have used up their gas more slowly than the Galaxy in the past, at present they are passing through a phase of star formation which is relatively much more brilliant than that in the Galaxy. There are galaxies which are more extreme examples of irregular star formation than the Magellanic Clouds. These include *starburst galaxies*, which at present have an intense burst of star formation which could not have been sustained throughout past galactic history. There are also the low mass *blue compact galaxies* which have very low heavy element abundance but which are also currently undergoing an intense burst of star formation. I have assumed that star formation is a smooth function of gas density. The existence of these unusual galaxies shows that this cannot always be the case.

Summary of Chapter 7

This chapter has been concerned with the way in which the fractional mass and the chemical composition of the interstellar gas in galaxies vary during their lifetime. Most of the detailed discussion has been concerned with the Galaxy, although some brief remarks have been made about the properties of other galaxies. The amount of gas in a galaxy changes because star formation turns gas into stars and because mass is lost by some types of star. In addition there may be an exchange of gas between a galaxy and the intergalactic medium. The chemical composition of the gas changes because the mass ejected by stars is characteristically richer in heavy elements than the gas out of which the stars formed and than any accreted gas. A full study of the chemical evolution of galaxies requires knowledge of the processes of galaxy and star formation, the nuclear evolution which occurs inside stars of different masses, the degree of mass loss from stars and when it occurs, the manner in which mass lost from stars mixes with the already existing interstellar medium and the extent to which gas is exchanged with the intergalactic medium

Much of the chapter is concerned with a simple model of chemical evolution in the solar neighbourhood of the Galaxy. In this model it is assumed that the gas in the solar neighbourhood has uniform chemical composition and that the solar neighbourhood can be regarded as isolated both from the remainder of the Galaxy and from the intergalactic medium. The process of star formation is supposed to depend only on the density of the gas and it is also assumed that the only important mass loss occurs from stars which are so massive that their evolution time is very much less than the timescale of galactic evolution. This simple model is found to be totally incapable of explaining the observations relating to the solar neighbourhood. In particular it predicts the existence of many more low mass stars which are highly deficient in heavy elements than are observed.

The remainder of the chapter is concerned with modifications to the simple theory which may explain not only the observations of the solar neighbourhood but also observations of other parts of the Galaxy and of other galaxies. These modifications include the idea of prompt initial enrichment which occurs, for example, if the earliest generations of stars were preferentially high mass stars which produce significant quantities of heavy elements very rapidly, the suggestion that stars form most effectively in regions of gas with higher than average heavy element content (metal-enhanced star formation) and a study of the effect of accretion of gas by galaxies (infall). Infall is automatically relevant if the formation of stars starts before a galaxy has reached its final mass and it is generally believed that this allowance for the time involved in galaxy formation is the most important modification required to the simple model. It appears that infall may have been very important in the solar neighbourhood for most of past galactic history.

Both star formation and galaxy formation are not very well understood at the moment. As a result, *a definite discussion of the chemical evolution of galaxies is not possible but the basic ideas of the subject are becoming clear.*

8

Galaxies and the Universe

Introduction

It is not really possible to consider the formation and early evolution of galaxies without also considering *cosmology*, that is the structure and evolution of the whole Universe. The reason for this is that, as I have mentioned earlier in Chapter 3, we have no clear evidence for more than one epoch of galaxy formation and that epoch appears to have been shortly after the *origin of the Universe*, if we believe that the Doppler shifts in the spectra of distant galaxies indicate expansion of the Universe from an initially dense state. I am supposing that the galaxies have formed out of intergalactic (or more accurately pregalactic) gas in much the same ways as stars have formed out of the interstellar gas. However, there is one important difference. Once a galaxy has formed, its distance from other remote galaxies increases because of the expansion of the Universe, but there is no reason for believing that the galaxy itself expands. Thus, whereas star formation can be assumed to take place in a system of gas which is stationary apart from internal motions within a galaxy, the formation of a galaxy takes place against a background of general expansion of the pregalactic gas.

It is important to know when condensations of galactic size were first established, because this influences how much gravitational energy was released in galaxy formation. At present large galaxies are typically between 10 and 100 times as far apart as the largest linear dimension of their clearly visible parts. They must be effectively closer when account is taken of invisible matter. If their formation occurred quite recently in the history of the Universe, and I shall explain this phrase more carefully shortly, they will have contracted from several or many times their present size and there will have been a substantial gravitational energy release in galaxy formation. If, in contrast, protogalactic condensations existed even at the earliest epochs, the release of gravitational energy in the phases of galaxy formation immediately preceding star formation will be correspondingly less. In this case the protogalaxy might even have been initially smaller than its present size, although as we shall see on page 185 it must subsequently have become at least twice its present size. A related question of interest concerns the *efficiency of galaxy formation*. Did the process of galaxy formation lead to a situation in which essentially all of the matter of the Universe is inside galaxies or is there a significant amount of intergalactic matter today?

The hot big bang cosmological theory

Because this is not a textbook on cosmology, I shall not attempt to discuss galaxy formation and evolution and the interpretation of observations of galaxies in terms of a variety of cosmological models. Instead I shall only discuss things in the framework of one model, the *hot big bang cosmological model*, which appears to give a very good explanation of the few observations which we possess which are of undoubted cosmological importance. Although this model is currently very successful, there are some detailed problems in the comparison between theory and observation which may yet lead to modifications in the model.

The theory is one in which the large scale structure of the Universe is governed by the *general theory of relativity*. In this particular theory the Universe, on the average, is both *homogeneous* and *isotropic*, which means that its gross properties are the same at all points and in all directions at a given point. Although in such a relativistic theory it is not in general true that observers at different points will agree on measurements of space and time, it is possible to introduce a *cosmic time* for all observers locally at rest in the expanding Universe. In terms of this cosmic time, t, the distance between objects, which are taking part in the expansion of the Universe, can be expressed in terms of a scale factor $R(t)$, which is a dimensionless measure of the distance between them at time t.

The hot big bang theory is not a unique theory. In its simplest form, which is the one which I shall describe here, there is one free parameter. In all versions of the theory, both the temperature and the density were infinite at time $t = 0$. This is of course a mathematical idealisation. Provided the initial temperature was significantly greater than 10^{10} K, the subsequent behaviour of the homogeneous isotropic Universe is not crucially dependent on the precise form of this initial condition. However, the one apparently free parameter in the theory may, in fact, be determined by what happened at higher temperatures as may be departures from strict homogeneity and isotropy in the real Universe. Subsequently, for as long as the Universe is sufficiently homogeneous to possess a single value of both temperature and density, these quantities decrease. The different variants of the theory can be characterised by the value of the matter density at a given temperature shortly after the start of the expansion, say 10^{10} K. Note that according to the special theory of relativity all forms of energy possess a mass equivalent and the density which influences the expansion of the Universe includes the mass equivalent of kinetic energy as well as the mass of particles. For different values of this density, the scale factor $R(t)$ varies with time in one of the three ways shown in fig. 88. In case (a), $R(t)$ increases without limit with increasing

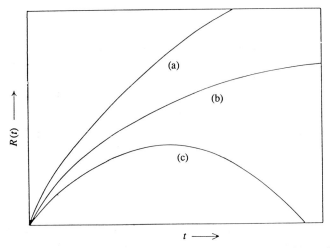

Figure 88.   The possible behaviour of the cosmological scale factor $R(t)$ with time.

t. The Universe† is then said to be *open*. In case (b), although $R(t)$ is unbounded as $t \to \infty$, $\dot{R}(t)$ asymptotically approaches zero. The Universe is again open. In case (c), $R(t)$ increases to a maximum value and then decreases again becoming zero at a finite value of t. In the last case the Universe is said to be *closed* and it has a finite volume at all t. Case (c) may lead to an *oscillating* Universe in which the first cycle is followed by other similar cycles, although no fully consistent theory of this type has yet been produced. It does, however, seem clear that the *bounce* must occur at very high temperature and density if it does occur at all, so that from the observational point of view an oscillating Universe may not look very different from a single cycle closed Universe. In any cycle the Universe may possess very little *memory* of the previous cycles.

I shall wish to discuss below what observations of galaxies, as well as related observations and theoretical ideas, can tell us about the closure or otherwise of the Universe. The models (a), (b) and (c) represent a series of models of increasing mean density at a given epoch because, the higher the mean density, the greater the probability that the mutual gravitational attraction of the material will be able to reverse the initial expansion. The critical density corresponding to case (b) at the present epoch depends on the value of the Hubble constant, the determination of which has already been discussed in Chapter 1. If the critical density is called ρ_{crit} we have

$$\rho_{\text{crit}} = 1.9 \times 10^{-30} H_0^2 \text{ kg m}^{-3}, \tag{8.1}$$

where the units of the Hubble constant are km s^{-1} Mpc^{-1}. In what ways can observations of galaxies tell us whether the mean density of the Universe is greater than or less than the critical density?

† In this section I refer to a variety of types of Universe. Of course there is only one Universe by definition. It is, however, inconvenient to have to talk about possible models of the Universe rather than the Universe.

The mean density of matter in the Universe

There are two different approaches to determination of the mean density of the Universe using observations of galaxies. The first is to ask what contribution the total mass of all known galaxies makes to the critical density. This means that we add up the total mass in galaxies in a given volume and divide by the volume to get a mean density. I have already stressed the uncertainties in galactic masses in Chapters 3 and 5, but the result of this calculation, using the most generally accepted values of galactic masses, is that the mass in galaxies does not go very far towards closing the Universe. Thus the average smoothed out density of galactic matter is estimated to be

$$\rho_{gal} \approx 2 \times 10^{-32} H_0^2 \text{ kg m}^{-3}, \tag{8.2}$$

or about one per cent of the critical density. Note particularly that Hubble's constant enters into (8.1) and (8.2) in the same manner so that the large uncertainty in the value of H_0 does not affect the estimate of ρ_{crit}. This value of ρ_{gal} includes the known luminous and interstellar matter in galaxies but, as I have said in Chapters 3 and 5, it may be significantly less than the true value of ρ_{gal}. If, as seems essentially certain, some galaxies have massive haloes, this will go some way to reducing the difference between ρ_{gal} and ρ_{crit}, but it is difficult to believe that there is enough mass in galaxies to close the Universe. This suggests that unless there exists a substantial amount of matter outside galaxies, the Universe is open. Unfortunately, there is at present neither a direct observation of the presence of a general intergalactic medium nor a conclusive proof that it does not exist, so that we cannot make a straightforward allowance for the amount of intergalactic matter. Hot gas emitting X-rays is observed in some clusters of galaxies but its total mass does not make a substantial difference to the amount of matter known and it is not enough to explain the virial mass discrepancy in clusters to be described shortly.

The second approach to the problem involves using observations of galaxies to obtain indirect information about the presence of intergalactic matter, which may close the Universe. There are two different types of observation. The first depends on the observations of clusters of galaxies which I have described in Chapters 3 and 5. As I have shown there, the total mass of many galaxy clusters, obtained from the Virial Theorem using the assumption that they are gravitationally bound systems, is much greater than the mass deduced from the estimates of masses of individual galaxies in the clusters. This indicates one of four things: that the masses of individual galaxies *are* very much higher than has been estimated or that there is a large amount of intergalactic matter inside clusters or that the clusters are chance superpositions on the sky and are not gravitationally bound systems or that the clusters are still forming so that their kinetic energy exceeds the virial value and some clusters may not yet be in equilibrium. Although there are undoubtedly some cases of chance superpositions being mistaken for clusters, there seems no doubt that the *virial mass discrepancy* is real in some cases and the presence of intergalactic matter would be the most attractive way of solving it.

Estimates of the total amount of matter in clusters of galaxies suggest that they may contribute between $0.1\rho_{crit}$ and $0.2\rho_{crit}$ to the mean density of the Universe. This also suggests that the Universe is open.

The other observation is related to the expansion of the Universe and to the establishment of the distance scale in the Universe which has been discussed briefly in Chapter 1. From fig. 88 it can be seen that, whether or not the Universe is closed, the expansion should be slowing down if the big bang theory in its simplest form is valid. This slowing down can be expressed in terms of a *deceleration parameter* which at the present epoch has the value

$$q_0 = -\ddot{R}_0 R_0 / \dot{R}_0^2, \tag{8.3}$$

where R_0, $\dot{R}_0$ and $\ddot{R}_0$ are the values of R and its first two time derivatives at the present epoch. If the Universe is just closed, solution of the equations gives $q_0 = 0.5$. If q_0 is larger, the Universe is closed, if it is smaller, it is open. Can a value of q_0 be obtained from observation?

Determination of deceleration parameter

We could obtain q_0 from observations, if it were indeed true that we have *standard candles* which we can use at really great distances. Thus, when we observe nearby galaxies, we find that there is a linear relation between their velocity of recession deduced from their redshift and their distance as deduced from the application of the standard candle techniques. As we observe objects at greater and greater distances (or equivalently redshifts) there are deviations from the linear law

$$v = H_0 r. \tag{8.4}$$

In principle, observations of deviations from this law can be used to obtain a value of the deceleration parameter and hence of the present mean density of the Universe, since the deviations from the straight line plot depend on the deceleration parameter (fig. 89). The relationship between apparent luminosity and redshift, which is equivalent to the relation between distance and velocity is here shown for several values of the deceleration parameter. I can now ask which of such a set of curves gives the best agreement with observations.

In fact there are difficulties in this method for two different types of reason. The first is just that it is difficult to obtain enough really reliable observations of extremely distant objects, although modern telescopes and ancillary equipment which is much more efficient than the photographic plate are gradually improving this situation. The largest observed galaxy redshift is now in excess of 3, where redshift is defined by

$$z = \delta\lambda/\lambda, \tag{8.5}$$

with λ the laboratory wavelength of a spectral line and $\delta\lambda$ its shift to the red. It can be seen from fig. 89 that it should in principle be possible to determine q_0 quite accurately from observations of standard candles at redshifts of unity and above.

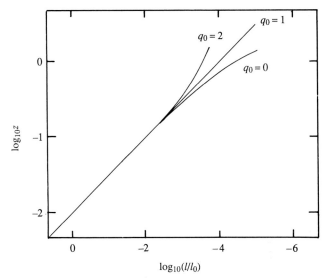

Figure 89. A plot of apparent luminosity (*l*) against redshift (*z*) for a standard candle for three values of the deceleration parameter q_0. l_0 is a normalising luminosity.

The second difficulty is concerned with the whole concept of standard candles. If the properties of the standard candle (for example the luminosity of a giant elliptical galaxy or the diameter of a cluster of galaxies of given richness) are not really constant but are themselves varying with time, what I may be obtaining is not the deceleration parameter but the variation of the standard candle, or more precisely some combination of the two things. It is then necessary to have some theoretical ideas about how the standard candle may have changed.

Variations of galactic luminosity with galactic age

Most observations used to date have been apparent luminosity/redshift relations for the brightest galaxies in rich clusters of galaxies. When we observe very distant galaxies, we are observing them as they were a long time in the past. Is it reasonable to suppose that they had the same luminosities then that giant elliptical galaxies have now in our neighbourhood? To answer this question I must consider how the luminosity of a galaxy is likely to vary during its life history. In principle, I use models of galactic evolution of the type which I have discussed in the previous chapter and add up the total luminosity of all the stars that are present at any given time, allowing for the variation in luminosity of a given star as it goes through its evolution. In practice, as we have seen, it is probably sufficient to assume that an elliptical galaxy has only one important generation of stars and to calculate how the total luminosity of that generation changes with time.

At first sight one might expect that the luminosity of a galaxy would be very much greater when it is young than when it is about as old as the Galaxy because the massive luminous stars are only present at that time. The actual situation is not

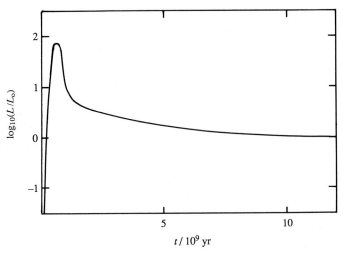

Figure 90.   The schematic time variation of the luminosity of a giant elliptical galaxy, with L_0 being the present luminosity.

quite as extreme as this because the less massive stars increase substantially in luminosity when they become red giants and this compensates to some extent for the loss of the more massive stars. However, when all these factors are taken into account it does appear that galaxies are more luminous when they are young, as is shown in fig. 90. This initial high luminosity is characteristic of the giant elliptical galaxies which are used as standard candles. If this effect is allowed for, distant galaxies are further away than if they were true standard candles and the apparent value of the deceleration parameter must be adjusted. The effect is that the true deceleration parameter is less than the apparent one. When corrections due to variations of galactic luminosity with time were first discussed it appeared that the Universe was almost certainly open. At one time there was even a suggestion that the observations could only be understood with a negative value of q_0. Such a value would imply that the expansion of the Universe was speeding up rather than slowing down. Such a result could not be understood in terms of the simple big bang theories which I am describing in which gravitational attraction must produce deceleration. To explain such results there would need to be an additional repulsive force driving the expansion of the Universe. Such a repulsive force associated with what is known as a cosmological constant was suggested by Einstein as a modification to general relativity before the expansion of the Universe was discovered. Einstein believed that the Universe must be static and he needed the repulsive force to prevent the collapse of the Universe because of its gravitational attraction.

The influence of galactic collisions

More recently it has been realised that there is a further factor which may affect the luminosities of the most massive galaxies in clusters and that this factor would cause their luminosity to increase with time. The mechanism is as follows. Galaxies in clusters move in the gravitational field of all the other galaxies in the same way as stars in a single galaxy interact as described in Chapter 4. In Chapter 4 I have argued that stars at average positions in galaxies do not very frequently collide with one another. Indeed I have said specifically that the Sun is unlikely to collide with another star until the Galaxy is very much older than it is now. The main reason for this is that distances between stars are very much larger than stellar radii. The same thing is certainly not true of galaxies and in particular of galaxies in rich clusters. Near the centres of rich clusters direct collisions between galaxies can have occurred in significant numbers during the present age of the Universe. The effect must be even greater if galaxies do possess massive haloes. The most important feature of such collisions for our present discussion is that the most massive galaxy in a cluster will be at or near to its centre of mass and it will tend to accrete and swallow smaller galaxies that collide with it. We thus have a situation in which the most massive galaxies in clusters will get more massive and possibly more luminous as time passes. There is also an interesting possibility that the gas which appears to be important in the explosions which produce radio galaxies may come not from normal mass loss from stars but from an accreted galaxy; either directly as gas or through the disruption of some of its constituents by strong tidal gravitational forces during the collision. There is also evidence that interactions between galaxies and mergers may stimulate an increased rate of star formation. Calculations of the rate of growth of mass of giant galaxies suggest that it could be several per cent in 10^9 years, so that a substantial increase of mass could occur in the galactic lifetime. At present the value of the consequent increase in luminosity is even less certain and it is not possible to say whether this effect is sufficient to cancel out the change in luminosity due to the steady evolution of the galaxy, which I have previously described. Another way in which giant galaxies increase in mass is through the accretion of intergalactic gas. In some clusters *cooling flows* are observed. The X-ray emitting gas is cooling and falling on to the massive galaxy at the cluster centre. The rate of increase in the galactic mass can be quite substantial but there does not appear to be a corresponding increase in luminosity. If the gas is forming stars, they are preferentially of low mass. What is clear is that a considerable amount of both theoretical and observational work will be required before it can be decided whether the apparent luminosity/redshift relation can yield very useful information about the closure or otherwise of the Universe.

There is one further manner in which observations of galaxies can in principle provide information about the mean density of the Universe. Although galaxies move apart due to the expansion of the Universe, their velocities show deviations from this smooth expansion. Some of these peculiar velocities are random and can be thought of as thermal motions of the galaxies but in other cases ordered streaming motions are observed. For example the Local Group of Galaxies is moving with a velocity of about 600 km s^{-1} relative to the local rest frame, which can be defined as that frame in which the cosmic microwave radiation, to be defined shortly, is isotropic. This motion and other streaming motions may be

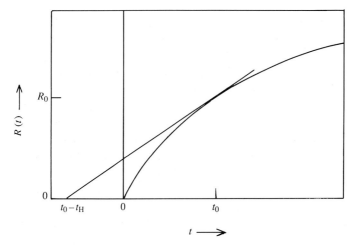

Figure 91   Relation between the age of the Universe, t_0 and the Hubble time, t_H.

caused by the gravitational attraction of a large mass concentration. It can then be asked what contribution such a mass would make to the mean density of the Universe. Several such investigations have suggested that the mean density could be significantly above the cluster virial value but probably still below the closure density. It has, however, to be stressed that, even if there is no flaw in such a discussion, it will only give a lower limit to the mean density because an additional smooth distribution of dark matter could be added without changing the observations.

Other evidence bearing on the closure of the Universe

Because this is a book on galaxies rather than on cosmology, I will not discuss in detail other arguments which bear on this question, but for completeness it is necessary to mention them. They are of three types. The first involves comparing the ages of astronomical objects with the time available since the origin of the Universe. The second involves some detailed consideration of the chemical composition of the Universe. The third concerns theoretical ideas about the very early stages of the expansion of the Universe. It can be seen from fig. 88, and in more detail from fig. 91, that the time since the big bang is necessarily less than the Hubble time defined by

$$t_H = H_0^{-1}, \tag{8.6}$$

which is estimated from the present rate of expansion of the Universe. The more open the Universe is, the closer is the age of the Universe to t_H. The present best value of H_0 is, as we have seen, about 50 km s^{-1} Mpc^{-1} which leads to $t_H \approx 2 \times 10^{10}$ yr, although some workers argue strongly for a value around 75 km s^{-1} Mpc^{-1} and $t_H \approx 13 \times 10^9$ yr and there is also a group which favours 100 km s^{-1} Mpc^{-1} and $t_H \approx 10^{10}$ yr. The age of the Universe is necessarily less than

t_H and the oldest objects of which we know must be younger than this if the theory is to be consistent. Their actual ages can in principle be used to give an upper limit to the value of q_0. Theoretical calculations of stellar evolution combined with observations of the shape of the Hertzsprung–Russell diagrams of globular star clusters enable an estimate to be made of the ages of the clusters. A typical value found is 16×10^9 yr, with an uncertainty of 3×10^9 yr, but some investigators find values which differ from this by $\pm 4 \times 10^9$ yr. It is clear that the estimated ages of globular clusters argue for a low value of H_0 and even then suggest an open rather than a closed Universe. Attempts have also been made to estimate the age of the disk of the Galaxy by studies of the abundances of heavy radioactive elements and their decay products in the solar system. These elements are supposed to have been produced by nuclear reactions in stars and to have then decayed freely. The present observed abundances can be extrapolated back to what they were when the Solar System formed 4.6×10^9 yr ago. It is then possible to ask what production pattern over what timescale is needed to give the primeval Solar System abundances. This discussion has led to an age of the galactic disk of about 12×10^9 yr but there are very large uncertainties in this value. For consistency the age of the disk should be somewhat less than that of the globular clusters.

The second consideration relates to the abundance of helium and deuterium in the Universe. If it is supposed that these were produced by nuclear fusion reactions in the early stages of the hot big bang Universe, the observed abundance of ^{4}He and the deduced primeval abundance of ^{2}H give information about the density of the Universe at the time that the nuclear reactions were occurring. The abundances of ^{2}H and ^{4}He, as well as ^{3}He and ^{7}Li, depend on the number of neutrons and protons per unit volume at the time that nuclear reactions occur. These numbers are easily related to the present density of the Universe in the form of ordinary matter. The abundances also depend on two other factors: the rate of the weak interactions, which convert neutrons into protons, and the number of types of neutrino or other weakly interacting particles, which exist in equilibrium with radiation at high temperatures. The more types of neutrino there are, the more rapidly the Universe expands and cools. These two factors determine the relative number of protons and neutrons when the nuclear interactions take place. As essentially all of the neutrons and an equal number of protons form ^{4}He, this determines its abundance, which depends only slightly on the matter density. In contrast the abundances of ^{2}H, ^{3}He and ^{7}Li depend very strongly on the density. Experiments at CERN suggest strongly that there are only three types of neutrino (associated with the electron, muon and tauon) and the half life of the neutron, which is related to the rate of the weak interactions, appears from recent measurements to be close to 10.2 min. The pregalactic abundances of the light elements can be estimated from observations and they are consistent with the theoretical predictions, using the above neutron half life and three neutrino types, provided that the density is substantially below the closure density. The precise value depends on the value of H_0 but is probably rather more than the mass in galaxies (8.2) but less than the Virial Theorem estimate.

Does this mean that there is an inconsistency between theory and observation?

It would do if the only form of mass in the Universe today was ordinary matter. It is however possible that much if not most of the mass in the Universe is in the form of weakly interacting elementary particles. In the big bang theory any particle of mass m_P and its antiparticle exist in equilibrium with radiation when $m_P c^2 \lesssim kT$. At lower temperatures the particle/antiparticle pairs annihilate unless their interactions are so weak that their mean free path for annihilation exceeds the radius of the observable Universe. For neutrinos this happens when $T \approx 10^{10}$ K. Subsequently, the particles are said to have decoupled and, if they are stable, they remain in the Universe today. If such particles have a non-zero rest mass, they may represent the main form of mass in the Universe. One possibility is that neutrinos have a small mass, which is not ruled out by current experiments. Another possibility is that some other weakly interacting particle exists which is stable and has a finite mass. Suggestions relating to such particles have been made on the basis of theoretical models of the properties of elementary particles, but none has yet been detected experimentally.

There have been theoretical suggestions that the value of $\Omega_0 \equiv \rho_0/\rho_{crit}$, where ρ_0 is the present total mean density of the Universe, should be very close to unity. The suggestions are related to two problems known as the *horizon problem* and the *flatness problem*. One of the principal observations in favour of the hot big bang cosmological theory is that of the *cosmic microwave background radiation*. This is radiation reaching the Earth from all directions which is essentially black body with a temperature of 2.7 K. It is interpreted as radiation from the early stages of the expanding Universe, which decoupled from matter, in the sense that its mean free path for absorption or scattering exceeded the radius of the observable Universe, when the temperature was a few times 10^3 K and which has expanded and cooled ever since. In fact the value of the present temperature of the radiation is required to relate the present density of matter to that when the nuclear reactions occurred because as the Universe expands $\rho \propto T^3$. The microwave radiation is remarkably isotropic. Its intensity is the same at opposite sides of the sky although these regions have never been in causal contact, in the sense that a signal could not have passed from one to the other in the present age of the Universe. Unless we are prepared to accept that the initial conditions at the start of the expansion of the Universe were ones of precise isotropy, we need to understand this horizon problem. The flatness problem concerns the variation with time of the ratio of the actual density to the closure density. If this is called Ω, it is possible to show that for most of the past history of the Universe there is a simple relation

$$\Omega = \Omega_0(1 + z)/(1 + \Omega_0 z) \tag{8.7}$$

between Ω, Ω_0 and the redshift, z, of an object whose radiation would just be reaching us now, if it was emitted when the density parameter had the value Ω. It can be seen that, for large values of z, $\Omega \approx 1$ whatever is the value of Ω_0. $\Omega \equiv 1$ is the flat Universe and it can be seen from (8.6) that the initial Universe must have been essentially flat whether $\Omega_0 \approx 0.01$ or 1.0. The microwave radiation is reaching us from $z \approx 10^3$.

It has been suggested that in the extremely early Universe ($t \approx 10^{-35}$ s; nuclear reactions occur when $t \approx 10^2$ s) matter has a strange equation of state in which it exerts a tension rather than a pressure. When this equation of state is inserted in Einstein's equations, it predicts an exponential expansion of the Universe. Because this expansion can be much faster than the speed of light, regions which were once in causal contact can get out of contact. In addition a large expansion can make curved space effectively flat. This *inflationary* phase may possibly explain both the horizon problem and the flatness problem, although one consequence of this explanation is that we may live in a smooth bubble in an infinite Universe whose properties are quite different well outside our observable Universe. Another consequence is that we might expect to find that Ω_0 is essentially equal to unity. If that is the case there seems little doubt that much of the matter in the Universe must be in the form of weakly interacting particles. There is not at present complete agreement that the Universe did have an inflationary phase but it is a subject of very active research. The expansion of the Universe at a speed greater than that of light does not contradict the special theory of relativity. This states that no information must travel faster than light and the inflationary Universe model preserves this property.

At present it appears that there is a reasonable understanding of the observed properties of the Universe in terms of the standard hot big bang cosmological theory although it is impossible to state at present whether the Universe is open or closed. It must of course be stressed that we do not *know* that we live in the hot big bang Universe. All the statements that have been made above are based on the assumption that the hot big bang theory is valid. *This may not be true.* In particular, if any future observation provides an irreconcilable contradiction with the predictions of the theory, this will imply that the standard hot big bang cosmological theory is invalid. I shall shortly mention one possible problem. Note that the standard theory predicts the relation

$$\Omega_0 = 2q_0. \tag{8.8}$$

If an accurate value of q_0 can eventually be measured it may put a constraint on Ω_0.

The formation and early evolution of galaxies

Having given above a somewhat inconclusive discussion of how observations of galaxies can give information about the past and future behaviour of the Universe, I return to the first point which was mentioned in this chapter, the problem of the formation and early evolution of galaxies. Let us first ask an observational question. Is there any possibility that we can see far enough back in the past to observe newly forming galaxies and is there even any possibility that galaxies are still being formed today? The most distant objects which have been observed at present are the quasars which frequently have redshifts in the range $z = 2$ to $z = 4$, with the highest observed value of z currently being 4.9. This assumes that the quasar redshifts do arise from the expansion of the Universe so that they do indicate distance. It is *possible* that the main epoch of galaxy formation was the same epoch in which quasars were common and, in fact, many people believe that a quasar is nothing more than the extremely luminous nucleus of a forming galaxy. A quasar appears as a point source of radiation and certainly most of its luminosity comes from a volume very much smaller than an ordinary galaxy, but it could easily be surrounded by a faint halo of galactic dimensions. Some quasars are surrounded by a more extended region of faint emission and others are associated with clusters of galaxies. Some galaxies have now been found with $z > 3$.

Redshifts as a measure of the passage of time

According to the big bang theory of the Universe, the redshift of a distant galaxy or quasar is directly related to its distance from us, although the conversion of redshift to distance requires a value of the deceleration parameter.† In addition we are observing these distant objects as they were in the remote past and the time in the evolution of the Universe at which they are being observed is also directly related to the redshift. Because of this it is possible to use z as a measure of passage of time in the evolution of the Universe. Thus we can, for example, say that we know that quasars were active at redshift $z = 4$ and ask whether the most important time for galaxy formation was at higher or lower z than this. The advantage of using z to measure the passage of time is that it is directly observable. It can of course only be converted to give a real measure of time when q_0 is known. In what follows I shall use z in this manner. If $\Omega_0 = 1$, the relation between z and t is

$$t = t_0/(1 + z)^{3/2} \tag{8.9}$$

where t_0 is the present age of the Universe. If this is the case the radiation from the quasar with $z = 4.9$ left it when the Universe was about seven per cent of its present age.

I now return to the question of when galaxies formed. Leaving on one side the possibility that galaxies and quasars may be very closely related, what constraints can be placed on the epoch of galaxy formation? I am supposing that galaxies form by a gravitational condensation out of the pregalactic medium. If so they must certainly be smaller today than they were when the formation started. This means that the epoch of galaxy formation, by which I mean the final stage of rapid collapse from a protogalaxy to form a recognisable galaxy, must have been after the time when the mean density of the Universe was equal to the present mean density of matter in galaxies. The value of the latter depends on whether or not the typical galaxy has a massive halo. At the largest the mean density of matter in galaxies is of order 10^6 times the mean density of the Universe today, but with larger galaxies and massive haloes and with any substantial amount of intergalactic matter the value could be as low as 10^4. Because, according to the big bang

---

† There is not even really a unique definition of distance.

theory, the density of matter in the Universe scales with redshift according to the law

$$\rho \propto (1 + z)^3, \qquad (8.10)$$

this implies that the epoch of galaxy formation could not be earlier than a time corresponding to redshift between 20 and 100.

The values of redshift that I have just quoted are considerably larger than the observed values for quasars, but there are further arguments suggesting that the actual epoch of galaxy formation must be considerably later than the earliest period just mentioned. The first point is quite a simple one. If a protogalaxy (or a protostar) collapses under gravity it can only come to equilibrium at a smaller radius if it loses energy. If its energy is conserved it is capable of re-expanding to its original size. We know from the Virial Theorem discussed in Appendix 2 that the total energy of a self-gravitating system in equilibrium is negative and is equal to half the gravitational potential energy. For a spherical system it is then easy to see that the radius of the initial equilibrium state must be less than or equal to half that of the initial condensation which is on the verge of collapse. Thus, if $r_{initial}$ and r_{final} are the initial and final radii, the initial energy is $-\alpha GM^2/r_{initial}$, where α is a constant of order unity and the final energy is $-\alpha GM^2/2r_{final}$ (half final gravitational energy). As energy must have been lost, $r_{final} \leq r_{initial}/2$ follows. This implies that galaxy formation must have started when the mean density was less than 1/8 of the present mean density inside galaxies and this reduces the possible value of z at the epoch of galaxy formation by a factor of approximately 2.

The other points are slightly more complicated. Galaxies will only form if there is an initial fluctuation of density which is higher than the mean density of the Universe. As it is this enhanced density which must be compared with the present density inside galaxies through our previous argument, the epoch of galaxy formation must be placed a little later when the value of z is still lower. In addition, if the protogalaxy is to collapse significantly more rapidly than the Universe expands, so that the *period* of galaxy formation is not too long, this density enhancement must be fairly substantial, probably by an order of magnitude. If we make these assumptions, it becomes difficult for the epoch of galaxy formation to be earlier than $z = 10$ and it is probably later than that. It then appears to be not at all impossible that quasars and forming galaxies are related and that galaxies in formation can be observed. To put the redshift z and time in a slightly more precise relation, in a Universe which is just closed a redshift $z = 10$ corresponds to a time about 4×10^8 years after the origin, the precise value depending on H_0.

There are more detailed considerations which are inappropriate to the present book but which are crucially important. I have already stated that the ultimate size of a galaxy depends on how much energy is dissipated during its formation. The precise dissipation of energy needs to be studied in detail before the actual sizes and masses of galaxies can be understood. I should make a few comments on the *mechanism* of galaxy formation. If the Universe really was completely homogeneous in its earliest stages of expansion, how did the density irregularities arise which led to the protogalaxies? Here it must be stated that there is, as yet, no very convincing picture of galaxy formation. Small statistical fluctuations in a smooth expanding Universe do not grow rapidly enough to form galaxies when they are needed. It is therefore necessary to assume that the Universe possessed a spectrum of density fluctuations even at the earlier epochs.

The discussion of how the observed galaxy distribution can have resulted from an initial spectrum of fluctuations is a major field of research today. The fluctuations can be *isothermal* in which there are perturbations of matter density but not radiation or *adiabatic* in which the matter and radiation vary together. The ultimate behaviour of the fluctuations depends on what is the principal form of mass in the Universe. If it is low mass neutrinos, the first objects which form in the Universe are very massive corresponding to superclusters of galaxies and the observed galaxies then have to form by fragmentation. If the principal form of mass is more massive but as yet hypothetical elementary particles, the first objects are nearer to globular cluster size and the galaxies and clusters of galaxies must form by aggregation under the force of gravity. In either case numerical simulations are carried out to determine whether the observed distribution of galaxies, with its clustering properties, can arise by the present epoch. There is as yet no clear view about which is the best model. There are various complications such as the light distribution, which is observed, not necessarily following the mass distribution, which is determined by the calculations, and the possibility that explosions in the first generation of objects formed might influence subsequent evolution. On the observational side, until very recently there have been worries about the continuing failure to detect small scale anisotropies in the microwave radiation, which places constraints on fluctuations at $z \approx 10^3$. Now it has been announced that the COBE satellite has detected anisotropies which are of the correct order of magnitude to provide an understanding of how galaxies have formed. As this book goes to press this is a very active field of research.

The origin of galactic shape

If I suppose that protogalaxies do form and start collapsing to form galaxies, I have to explain why some galaxies (S, S0, SB and IrrI) are highly flattened, whereas E galaxies are only very slightly flattened. Here, once again, I should interpolate the remark that the true discrepancy will not be as great as the apparent discrepancy if galaxies do have massive haloes which are much less highly flattened than disks. I should also comment that high resolution photographs of galaxies have recently detected spiral features in some galaxies which were previously thought to be elliptical. An obvious explanation of the flattening of galaxies is that the highly flattened ones are rapidly rotating and that the ellipticals rotate much more slowly. We know that this is a good explanation of the shape of the disk of our Galaxy but the general explanation is not as simple as it appears at first sight.

If a rotating protogalaxy contracts and if its angular momentum is conserved, as it certainly will be if there is no influence such as a pregalactic magnetic field

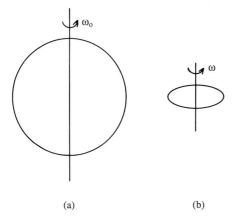

(a)   (b)

Figure 92.   The flattening of a rotating contracting cloud.

connecting the protogalaxy to its surroundings, it will necessarily rotate more rapidly and become flattened. Thus consider the collapse of a spherical cloud and assume initially that it remains spherical as it shrinks. If initially its angular velocity is ω_0 and its radius is r_0 and subsequently their values are ω and r, conservation of angular momentum implies that

$$\omega r^2 = \omega_0 r_0^2. \tag{8.11}$$

The gravitational force per unit mass towards the centre of the cloud is GM/r^2 where M is its mass. An element of the cloud on the equator would require a gravitational field or force per unit mass $\omega^2 r$ to keep it in circular orbit about the centre of the cloud. As the cloud contracts spherically the gravitational field becomes smaller than $\omega^2 r$ calculated using (8.11) and as a result the parts of the cloud at and near the equator move towards the axis more slowly than they would in spherical collapse and the cloud flattens. This is illustrated in fig. 92.

The flattening is not, however, enough to make certain that a flattened galaxy will be formed. The collapsing protogalaxy possesses enough energy to re-expand to its initial radius if energy is not dissipated when the system is in the highly contracted state. I have already explained on page 185 that some dissipation must occur if a bound galaxy is to form and that the final galaxy must have a size less than about half that of the original cloud. Such a size would not produce a very flattened galaxy from an initially approximately spherical cloud. Whether or not enough dissipation occurs to produce a highly flattened galaxy depends on the fraction of the system which is still a diffuse gas when it enters a flat state. If much of the system is still gaseous, collisions between gas particles will cause atoms to enter excited states and subsequently to radiate energy which can escape from the system. Such dissipation of energy can easily be sufficient to prevent the re-expansion of the disk. If much of the mass of the galaxy is already in the form of proto clusters and protostars, which are fairly compact, the probability of collision will be much reduced and there will be less dissipation and the system will re-expand to a radius which is not substantially less than its initial radius.

We are thus led to consider the possibility that elliptical galaxies are not highly flattened because the initial process of star formation was much more efficient in them than in spiral galaxies, rather than because they were initially much more slowly rotating. In the case of a flattened galaxy like our own, the globular star clusters are believed to have been sufficiently condensed out at the stage of initial collapse for them to continue to occupy a volume which is very much larger than the disk. This is not, however, a complete explanation of the difference between spiral and elliptical galaxies; it merely alters the question which has to be asked. We now need to determine what are the properties of a protogalactic cloud which ensure that it will fragment to form stars very efficiently. When this question is answered, it may indeed turn out that rotation *does* play a key role in the difference between elliptical and spiral galaxies. Thus it is possible that slow rotation is the factor which leads to rapid fragmentation but this has certainly not been proved. There is also at least a hint from the observed masses of galaxies, which differ from type to type, that mass might also play a role.

There is an alternative suggestion for the formation of the more massive elliptical galaxies. This is that they have formed as a result of the merger of a number of sub-units, all of which might independently have formed flattened galaxies. As long as the angular momenta of these sub-units are randomly oriented, the final galaxy has a low total angular momentum per unit mass and avoids significant flattening. In such a model merger must encourage star formation so that incipient gaseous disks are turned into stars more efficiently than in disk galaxies.

Final stages of galactic evolution

In the present chapter I have given a very schematic discussion of the origin of galaxies and in Chapter 7 I discussed their evolution, and particularly the evolution of our Galaxy, up to their present age. The one topic in galactic history on which I have not at present touched is the question of ultimate galactic structure. Here the answer once again depends on cosmology. If the Universe is open or only just closed, which at present seems very probable, galactic evolution will be able to proceed to its natural end-point. If, in contrast, the Universe is very strongly closed, the expansion of the Universe will cease and it will re-contract to a singularity. In this case the singularity may be reached before the galaxies have completed their normal evolution; if this is so, some very remarkable effects will

Final galactic evolution in an open Universe

Consider first the case of an open Universe. The lowest mass stars in a galaxy, which have a phase in which nuclear reactions provide their energy release, have masses of about $0.1 M_\odot$ and luminosities (on the main sequence) of about $10^{-3}L_\odot$; they therefore have lifetimes which are about 100 times longer than the Sun or between 10^{12} and 10^{13} yr. There will therefore be a time of the latter order for which a galaxy will contain some ordinary luminous stars but eventually they will all fade into insignificance. When this has happened most of the galaxy's mass, excluding weakly interacting particles, will be in the form of dead stellar remnants, black dwarfs, neutron stars and black holes. At this stage in galactic evolution, stars in the solar neighbourhood in our Galaxy and at comparable positions in other galaxies will not have had many collisions with other stars, according to the estimate of time between collisions given in Chapter 4. Thus nothing very significant will have happened to the *dynamical* properties of most parts of galaxies. This will not be true of star clusters and the dense central regions of galaxies. Star clusters will largely have dispersed, probably leaving behind a relatively massive black hole composed of those stars which did not succeed in escaping. In addition, many stellar collisions, also leading to the formation of massive black holes, will have occurred in galactic nuclei. Also the process of galaxy swallowing in rich clusters of galaxies, which I have described earlier in this chapter, will probably have proceeded to completion.

In those galaxies that remain, there will now follow a very long period in which stellar collisions will gradually become important throughout a galaxy. These collisions will lead to the escape of some dead stars into the ever increasing volume of intergalactic space together with the gradual accumulation of others into more massive black holes. The galaxies will not be completely dead even at this stage. For example, a burst of luminosity might accompany the collisions of two stars and stars approaching a massive black hole may be tidally disrupted by it. The matter may form a disk around the black hole and further matter falling on the disk may emit X-rays in the same way as some X-ray sources are believed to be associated with black holes in the Galaxy today. Although the Universe would not be completely black it would be very dim indeed. Subsequently black holes can themselves evaporate as a result of Hawking radiation, which occurs because energy cannot be confined for an arbitrarily long time in an arbitrarily small volume. Before that happens to stellar mass black holes, protons may decay, if current ideas in particle physics are correct.

Final galactic evolution in a closed Universe

In contrast consider the contracting phase of a strongly closed Universe. If we ignore the thermal properties of the stars, we can expect that, as the galaxies approach one another again and reach the stage that their separations are comparable with their radii, collisions between galaxies will become very important and that stars will be readily knocked out of galaxies and captured by others. In fact, the stage will be reached when there are no longer any galaxies and the Universe is composed mainly of stars, many of which are assumed still to be luminous objects. Still ignoring the thermal properties, we might expect this Universe of stars to persist until the separation of stars is comparable with their radii, so that the mean density of the Universe is about 10^3 kg m^{-3}. After this the individual stars will cease to exist.

The above discussion is completely false because the thermal properties are vitally important. In the contraction phase instead of the light from distant objects being redshifted it is blueshifted and this means that the radiation field is much more intense than is suggested by a simple calculation. Long before the stars approach one another as described in the last paragraph, the radiation density is sufficiently strong to heat stellar surfaces to temperatures more characteristic of stellar interiors. This implies that stars would be destroyed at this stage. Note that not only the radiation currently being emitted by stars would be blueshifted but also all of the radiation previously emitted, which had not previously been re-absorbed and the background radiation left over from the big bang.

Summary of Chapter 8

The origin and early evolution of galaxies cannot be studied without at the same time discussing cosmology. Observations other than those concerned with galaxies suggest that the hot big bang cosmological theory gives a good first approximation to the large scale structure and evolution of the Universe. It is therefore sensible to discuss how observations of distant galaxies might give further information about the validity of the cosmological theory and might determine a free parameter left in the theory and to try to understand how and when galaxies would have formed in a big bang Universe. The discussion of the first of these points given in the chapter indicates that it is at present not easy to avoid circular arguments. If galaxies had luminosities which did not change with time, observations of the apparent luminosity of distant galaxies would lead to a determination of the deceleration parameter or equivalently the mean density in the Universe. However, there are good reasons for believing that the luminosities of galaxies do change and it will probably be some time before it is clear whether observations are giving a value of the deceleration parameter or information about the early evolution of galaxies.

Some information about the mean density of the Universe can also be obtained by application of the Virial Theorem to clusters of galaxies and by studying galaxy streaming motions. From these discussions it appears that the Universe is open unless there is additional more uniformly distributed matter. A discussion of the formation of light elements in the early Universe suggests strongly that much of the matter in the Universe must be weakly interacting elementary particles, while theoretical considerations suggest that the Universe should be very close to critical density in which case it is almost certain that the mass of the Universe is dominated by elementary particles.

It was not possible to discuss galactic formation in detail in the present book. It was, however, possible to make three general points. The first is that, if galaxy formation is to occur in a hot big bang Universe there must be some departures from strict homogeneity and isotropy at early epochs. The second is that the main epoch of galaxy formation may well be no further in the past than that at which we are observing the most distant quasars so that forming galaxies may be discovered. The third is that, although rotation must play some role in determining whether or not galaxies are flattened, the stage at which significant star formation occurs in a protogalaxy is possibly even more important.

The final stages of galactic evolution are also related to cosmology because they depend on whether or not the Universe contracts after its present expansion. If there is no contraction, and that at present seems most likely, galaxies will eventually become non-luminous objects in which all of the mass is in the form of black holes and dead stars of low mass. At an even later stage black hole evaporation and proton decay may further modify the composition of the Universe. In contrast, if there is a re-contraction, galaxies will collide and lose their identity and be irradiated by intense blueshifted radiation from other galaxies (and the cosmic background radiation) before they complete their normal evolution.

9

Concluding remarks

In this book I have attempted to describe the ideas which are fundamental to an understanding of the structure and evolution of galaxies. There has not been very much emphasis on the presentation of precise numerical values for the properties of galaxies. It should be evident that one of the reasons for this is that there are at present very considerable detailed uncertainties in the subject. It should also be clear that the subject is developing very rapidly. In this chapter I will highlight some of those branches of the subject which are particularly controversial or where quick progress seems likely.

Galactic masses and composition

One problem which is fundamental to the whole of the book is that the masses and sizes of galaxies continue to be very uncertain. Essentially all estimates of galactic masses must be regarded as lower limits and large masses are probably hidden in extensive low density haloes. Observations of rotation curves of spiral galaxies, which remain flat at large radii rather than converging to the Keplerian law $v_{circ} \propto \tilde{\omega}^{-1/2}$, show that they contain much mass which is not contributing to the luminosity of the galaxies. As the rotation curves have not turned over at the largest radius at which observations can be made, it is not clear just how large the galaxies are and how much mass they contain. It is more difficult to make mass estimates for elliptical galaxies, but it seems possible that they also contain much hidden mass. A determination of the total masses of galaxies is important both for a proper understanding of the properties of the galaxies themselves and for a knowledge of the extent to which larger galactic masses affect the virial mass discrepancy in clusters of galaxies and contribute to the total mass density of the Universe.

The form of the hidden matter in galaxies is also of great interest. Although it could in principle be low luminosity stars, there is an increasing belief that much of the mass in galaxies may be in the form of weakly interacting elementary particles,

192

such as neutrinos if they have a small but finite mass. There are cosmological arguments which suggest that the elementary particles must exist and that all of the non-luminous matter in the Universe cannot be ordinary matter. If the elementary particles are present in galaxies, they may simply act as a source of an additional gravitational field acting on the stars and gas at the present time, but they may have played an important rôle in galaxy formation.

The size and mass of the Galaxy

The uncertainties which have just been mentioned extend to our knowledge of the Galaxy. Since the first version of this book was published, there has been a downward revision of the accepted values of the distance to the galactic centre R_0 and the local circular velocity $v_{\phi 0}$. This revision has led to a reduction in the estimated mass of the Galaxy within radius R_0. At the same time there is increasing evidence that the Galaxy possesses a massive halo and a large radius and a total mass which could be $10^{12} M_\odot$ rather than $10^{11} M_\odot$. It seems clear that the halo is not composed of ordinary luminous stars which could be detected as very high velocity stars in the solar neighbourhood. It would be much more difficult to detect a halo composed almost entirely of dead stars. Experiments are being mounted which might detect a local density of elementary particles, if they are the main source of galactic mass.

Observational developments

Large telescopes are being provided with ancillary equipment which is capable of detecting very much fainter galaxies and galaxies at higher redshift and new measuring machines can determine the positions and apparent luminosities of millions of galaxies. This enables detailed statistical studies to be made of the clustering of galaxies and of the large scale homogeneity and isotropy of the distribution of faint galaxies. Assuming a standard cosmological model, in which high redshift corresponds to the distant past, this allows the possibility of studying distant, and hence young, galaxies and clusters of galaxies with the aim of comparing their properties with nearby older galaxies and clusters. There have already been some unexpected discoveries including large voids in which there are very few galaxies and the suggestion that there is a large scale periodicity in the galaxy distribution. Although this book is really concerned with the properties of individual galaxies, it seems likely that the formation of individual galaxies cannot be discussed in isolation from a discussion of the clustering properties of galaxies, as I shall explain shortly.

Other important observational developments are concerned with the study of star forming regions in our own and other galaxies. The dense clouds in which stars form are most readily studied in the infrared and millimetre region of the electromagnetic spectrum and, as larger telescopes covering these parts of the spectrum have come into operation, much information about the star formation process has accumulated. The Infrared Astronomy Satellite (IRAS) has provided particularly valuable observations. As well as giving a detailed picture of star forming regions in our own Galaxy, it has identified a class of highly luminous starburst galaxies, which have a rate of star formation per unit mass which is much higher than in normal galaxies. Most, if not all, of the starburst galaxies are having a gravitational interaction with another galaxy. Although observations of star formation are accumulating, there is not as yet a good understanding of what gross properties determine the overall rate of star formation at a particular location in a galaxy. It is however this that is needed for a discussion of galactic evolution.

Cosmological problems

The formation of galaxies and clusters of galaxies can only be discussed in the context of a particular cosmological theory and the hot big bang theory currently appears to fit the observations. This cannot be said to be conclusively correct until there is an agreed value for Hubble's constant and it is clear that the age of the Universe is greater than the age of objects in it, particularly our Galaxy. The age of the Universe can only be made precise if a value can be obtained for the deceleration parameter, which is related to the actual mean density in the Universe. The value of the Hubble constant is still uncertain by about 50 per cent, although most determinations of it claim a very much higher accuracy than this. It is difficult to be confident about a value of the deceleration parameter being obtained by direct measurements, because it is hard to establish really reliable standard candles at large distances. Instead more direct attempts are made to study the mean density of the Universe. Studies of motions in clusters of galaxies suggest that the density is about a fifth of that necessary to close the Universe, while studies of the abundances of light chemical elements, supposed to have been produced by nuclear reactions in the early stages of the Universe, suggest that the amount of ordinary matter is much less than that. Purely theoretical arguments suggest that the Universe should have just about closure density. It is possible that theoretical calculation of the manner in which structure develops in the Universe might lead to a determination both of the mean density of the Universe and the form which it takes.

The origin of structure in the Universe

In the standard cosmological model, in which the Universe is strictly homogeneous and isotropic apart from statistical fluctuations, it is impossible for galaxies to form by the time at which they are observed to be present. It is therefore necessary to suppose that there was a spectrum of initial fluctuations in density in the Universe. These fluctuations may, in fact, have arisen in the postulated inflationary period in the early Universe, which was mentioned briefly in Chapter 8. The recent observations of small scale anisotropies in the cosmic microwave mediation provide information about departures from regularity in the Universe at $z \approx 10^3$. It is now necessary to ask both whether the magnitude and spectrum of these fluctuations can be explained by processes in the very early Universe and whether gravitational forces acting on the fluctuations can have produced the observed distribution of galaxies and clusters of galaxies by the present epoch.

The development of structure depends on the mean density of the Universe and on the nature of the main form of matter present. If the Universe is mainly made of elementary particles, there is a theoretical choice between light particles and heavy particles. I say that there is a theoretical choice because the heavy particles have not been shown to exist and the light particles (neutrinos) may not be massive enough to be of interest. In the case of light particles (hot dark matter), the first objects that form gravitationally are very massive corresponding to superclusters of galaxies and the smaller objects have to form by fragmentation. In the case of massive particles (cold dark matter), the first objects are more of globular cluster size and the larger objects must form by aggregation.

The formation of structure is studied by what is known as N body simulations. The matter in a large volume of the Universe is replaced by a large number of point masses which interact gravitationally as the Universe expands and the distribution of mass at the current epoch is compared with observations of galaxy clustering in the real Universe. Until recently it was believed that cold dark matter simulations gave best agreement with the observations but the more recent discovery of very large scale structure in the galaxy distribution has cast some doubt on this. The important point for the topic of the present book is that, if the Universe is composed of cold dark matter, galaxies may have formed by the aggregation of quite large sub-units, whereas, if the Universe is composed of hot dark matter, galaxies might have had a much smoother original structure. This in turn is clearly important in trying to understand the chemical evolution of galaxies.

The origin of galactic shape

Given that galaxies do form, it is then necessary to try to understand why some of them become spiral galaxies and some of them become elliptical galaxies. As has been stressed several times in the book, just how different these types of galaxy really are depends on whether or not spiral galaxies have massive haloes, which are not highly flattened. However, if the haloes are mainly composed of weakly interacting particles rather than low luminosity or dead stars, there remains a distinct difference in the distribution of ordinary matter. It is tempting to ascribe the difference to variations in the value of angular momentum per unit mass. It is, however, clear that, even if this does prove to be the key quantity, the difference of galactic type must arise at least partially from a variation in the efficiency of star formation, which may itself be influenced by the angular momentum. It is also possible that the most massive galaxies, which are ellipticals, may have been formed by the aggregation of smaller objects whose angular momenta were randomly distributed.

The chemical evolution of galaxies

The mention of star formation leads me to the topic of the chemical evolution of galaxies. In Chapter 7 I have presented a highly simplified discussion of how the present composition of the interstellar gas in a galaxy, and in particular our Galaxy, can be related to its initial composition, which is supposed to be that which was produced in the big bang. The details of this subject involve galaxy formation, star formation and stellar evolution and mass loss from stars, together with any processes moving gas around in a galaxy. Essentially all of these topics are highly uncertain. I have already mentioned that it is unclear whether galaxies form from a smooth distribution of gas or by the clustering of discrete sub-units. The relation of the average rate of star formation to the properties of the gas out of which stars form also requires considerable clarification. There do exist theoretical models of galactic chemical composition which give reasonable agreement with observed properties of our own and other galaxies. However it is by no means clear that the assumptions which have been made in the models will agree with more detailed observations when these become available.

Other outstanding problems

I have mentioned above just a few of the major topics of present research and I could have mentioned many others. For example, although it is generally agreed that the spiral structure of galaxies must be a wave pattern rather than a material structure, there is not as yet a conclusive theory of the origin and maintenance of spiral waves. Indeed, it remains possible that some spirals result from self-propagating star formation. It now seems clear that quasars are at cosmological distances and are related to the nuclei of galaxies, but there are still many problems associated with the energy output of quasars and the explosive events in radio galaxies and other active galaxies such as the Seyfert galaxies. The realisation that many elliptical galaxies are triaxial has led to a considerable effort to build self-consistent models of triaxial galaxies of the type discussed in Chapter 4. In trying to understand the past evolution of our Galaxy, it would be useful to know how the thickness of the galactic disk has changed with time. As explained in Chapter 6, this involves a complex interaction between gas, magnetic field and cosmic rays with the stars providing energy for the gas and producing cosmic rays. Above all, it must be stressed again that a cosmological theory is involved in an interpretation of many of the properties of galaxies. Although the hot big bang cosmological theory does at present appear to be in reasonable agreement with the large scale properties of the Universe, this may not be the finally accepted theory and this could affect our understanding of some galactic properties. *In any case the subject of the structure and evolution of galaxies will be one in which there are disagreements as well as exciting new developments for many years.*

Appendix 1

Some factors influencing stellar spectra

An absorption line is produced if electrons occupying a particular energy level in an atom or ion absorb radiation and move to a higher energy level. If the levels have energies E_1 and E_2 the absorbed radiation has frequency v given by

$$E_2 - E_1 = hv. \tag{A1.1}$$

Prominent absorption lines of a given chemical element will only be produced in the atmosphere of a star if three conditions are satisfied:

   (i) The atom is present,
  (ii) the element has energy levels with a spacing such that it can absorb radiation of a frequency which is present in significant quantities,
 (iii) the element is in the correct state of ionisation and excitation for the relevant lower energy levels to be occupied.

It is upon (ii) and (iii) that I will comment. I discuss (iii) first.

If conditions in the stellar atmosphere were those of *thermodynamic equilibrium*, both the radiation present and the state of excitation of atoms and ions would be determined by the temperature, T, of the atmosphere alone. The intensity of radiation would be given by the *Planck function*

$$I_v = B_v(T) \equiv (2hv^3/c^2)/[\exp(hv/kT) - 1]. \tag{A1.2}$$

If an atom (or ion) has energy levels E_r and E_s, in thermodynamic equilibrium the numbers n_r, n_s of atoms in the two states obey the *Boltzmann law*

$$n_r/n_s = \exp((E_s - E_r)/kT). \tag{A1.3}$$

Finally the numbers of atoms per unit volume in two successive states of ionisation n_i, n_{i+1}† are related to the electron density n_e by *Saha's equation*

---

† The quantities in (A1.3) should now be written more correctly n_{ir}, n_{is}.

$$\frac{n_{i+1}n_e}{n_i} = \left(\frac{2\pi mkT}{h^2}\right)^{3/2} \frac{2B_{i+1}}{B_i} \exp(-I_i/kT), \tag{A1.4}$$

where I_i is the energy required to remove one electron from the atom in the ith state of ionisation, m is the mass of the electron and B_i, B_{i+1}, which are called partition functions for the two states, depend on the electron energy levels in the two ions and the temperature. Saha's equation depends on both the chemical composition, because the electrons entering into n_e can be provided by ionisation of any of the element present, and on the density because the density enters quadratically in the numerator and only linearly in the denominator on the left hand side of (A1.4). It is easy to see that if the density is increased at fixed T, the degree of ionisation is reduced. However, because the temperature enters exponentially, the dependence on T is much more important.

Conditions in the atmospheres of stars depart from thermodynamic equilibrium but in many cases the departures are sufficiently small that the ideas given above are relevant; there would of course be no spectral lines if thermodynamic equilibrium were exact. Thus the occupancy of energy levels may be approximately determined by (A1.3) and (A1.4) and the photons which are available to be absorbed have frequencies close to the maximum of the Planck curve (A1.2). In particular the degree of ionisation in a stellar atmosphere does depend (weakly) on its density as well as its temperature. Because the surface densities of giant stars are lower than those of dwarfs, some chemical elements may be in a higher state of ionisation in giants than in dwarfs. As a result, even if a giant and dwarf have the same surface temperature and chemical composition, their spectra may be different. This fact has led to the development of *luminosity criteria* mentioned on page 22.

If an element has energy levels such that it can absorb (for example) visible radiation, the considerations which we just described will determine whether it is in the correct state to absorb the radiation. Un-ionised hydrogen can absorb visible radiation but not if it is in its ground state. Hydrogen absorption lines in the visible region of the spectrum are found neither in hot stars, where the hydrogen is ionised, nor in cold stars, where it is in the ground state. In contrast there are some elements which do not even possess spectral lines which are potentially observable in most stars. An example is boron which has proved very difficult to detect.

Appendix 2

The Virial Theorem

Consider a set of N point masses interacting through only their mutual gravitational attraction. Let the ith mass m_i be at position (x_i, y_i, z_i) at time t. Then its equations of motion under the attraction of all the other particles are

$$m_i\ddot{x}_i = \sum_{j\neq i} \frac{Gm_im_j(x_j-x_i)}{[(x_i-x_j)^2 + (y_i-y_j)^2 + (z_i-z_j)^2]^{3/2}} = \sum_{j\neq i} \frac{Gm_im_j(x_j - x_i)}{r_{ij}^3},$$

$$m_i\ddot{y}_i = \sum_{j\neq i} \frac{Gm_im_j(y_j-y_i)}{r_{ij}^3} \qquad\qquad\qquad (A2.1)$$

$$m_i\ddot{z}_i = \sum_{j\neq i} \frac{Gm_im_j(z_j-z_i)}{r_{ij}^3}$$

where the dot denotes differentiation with respect to time, the summation is over all of the other particles and r_{ij} is the distance between m_i and m_j. I now multiply the first of equations (A2.1) by x_i, the second by y_i and the third by z_i and sum over the equations for all of the particles.

Consider first the left hand side of the resulting equation.

$$\sum_i m_i(x_i\ddot{x}_i + y_i\ddot{y}_i + z_i\ddot{z}_i) = \frac{d}{dt} \sum_i m_i(x_i\dot{x}_i + y_i\dot{y}_i + z_i\dot{z}_i)$$
$$- \sum_i m(\dot{x}_i^2 + \dot{y}_i^2 + \dot{z}_i^2). \qquad (A2.2)$$

The second term on the right hand side of (A2.2) apart from the minus sign is $2T$, where T is the total kinetic energy of the system. The first term can be further rearranged as

$$\sum_i m_i(x_i\dot{x}_i + y_i\dot{y}_i + z_i\dot{z}_i) = \frac{1}{2}\frac{d}{dt} \sum_i m_i(x_i^2 + y_i^2 + z_i^2). \qquad (A2.3)$$

I define

$$I \equiv \sum_i m_i(x_i^2 + y_i^2 + z_i^2) \tag{A2.4}$$

and call it the *moment of inertia* of the system; it is half the sum of the three principal moments of inertia as they are usually defined.

Consider next the right hand side of the summed equation. For any two particles this will contain a term $Gm_j m_i x_i(x_j - x_i)/r_{ij}^3$ and a corresponding term $Gm_j m_i x_j(x_i - x_j)/r_{ij}^3$. These sum to $-Gm_i m_j(x_i - x_j)^2/r_{ij}^3$. When this term is added to the two similar terms involving y_i and y_j and z_i and z_j the numerator of the resulting expression contains r_{ij}^2 which can be cancelled with a similar term in the denominator. Thus the total contribution from the interaction between m_i and m_j becomes

$$-Gm_i m_j/r_{ij},$$

which is just the mutual gravitational energy of the two particles. The sum over all pairs of particles now gives

$$-\sum_{i,j\neq i} Gm_i m_j/r_{ij} \equiv \Omega, \tag{A2.5}$$

where Ω is the total gravitational potential energy of the system.

Using equations (A2.2) to (A2.5), the summed equation can be written

$$\frac{1}{2}\frac{d^2I}{dt^2} = 2T + \Omega. \tag{A2.6}$$

This is the general form of the *Virial Theorem* for a system whose properties may be changing with time. For a system whose overall properties are time-independent, the left hand side of equation (A2.6) must vanish and the Virial Theorem becomes

$$2T + \Omega = 0, \tag{A2.7}$$

which is an equation which is frequently used in this book. From (A2.7) it can be seen that the total energy, E, of the system is negative and satisfies

$$E \equiv T + \Omega = \Omega/2. \tag{A2.8}$$

Deductions for spherical systems

The gravitational potential energy of a spherical system of mass M and radius R is

$$\Omega = -\alpha GM^2/R, \tag{A2.9}$$

where the value of α depends on the mass distribution in the system but is usually of order unity. Also

$$T = \tfrac{1}{2}M{<}v^2{>},\tag{A2.10}$$

where $<>$ denotes an appropriate mean value. Combination of (A2.7), (A2.9) and (A2.10) for a spherical system in equilibrium, shows that the mean velocity of the particles is approximately the escape velocity defined by

$$v_{\mathrm{esc}}^2 = 2GM/R.\tag{A2.11}$$

If a spherical self-gravitating system contracts from rest and eventually reaches a state in which the Virial Theorem in the form (A2.7) is satisfied, its radius must then be less than or equal to half its initial radius. For, if no energy is lost, we can equate the initial energy of the system (which is entirely gravitational) to the final energy which is half the final gravitational energy. Thus

$$\alpha GM^2/R_{\mathrm{init}} = \tfrac{1}{2}\alpha GM^2/R_{\mathrm{final}}.\tag{A2.12}$$

In fact energy is probably lost so that the left-hand side is less than the right-hand side (as gravitational energy is negative). Thus

$$R_{\mathrm{final}} {\leqslant} \tfrac{1}{2}R_{\mathrm{init}}.\tag{A2.13}$$

(This is not a rigorous proof because α has not been proved to be constant but the result is true in most cases.)

Appendix 3

Gravitational fields due to spheres and ellipsoids

In this appendix I quote some results which are used in Chapter 5. The proof of most of them is outside the scope of the present book, and no proofs will be given, but they are discussed in standard books on the subject such as W. D. MacMillan, *The Theory of the Potential*, Dover Publications, New York. The results needed are the following:

1. If a mass distribution has spherical symmetry with its centre at the origin of spherical polar coordinates $r = 0$,
   (a) at a radius $r = r_1$ inside the body, the matter at radii outside r_1 exerts no net force and the field is the same as if all the mass interior to r_1 were at the origin,
   (b) at a radius $r = r_O$ outside the body, the field is the same as if all the mass were at $r = 0$.

2. If a body is composed of concentric ellipsoidal shells of the same eccentricity with density constant on any thin shell,
   (a) no net field is produced at a given point by the ellipsoidal shells outside that point,
   (b) far enough from the body (or indeed from a body of any shape or mass distribution), the field is the same as if all of the mass of the body is at its centre of mass.

3. If an oblate spheroid has uniform density, mass M_{sph}, semi-major axis a, eccentricity e and equation

$$(\bar{\omega}^2/a^2) + [z^2/a^2(1 - e^2)] = 1, \tag{A3.1}$$

   the gravitational field in its midplane is

$$g_{\bar{\omega}} = -(3GM_{sph}/2a^3e^3)(\beta - \sin \beta \cos \beta)\bar{\omega}. \tag{A3.2}$$

   where inside the spheroid

$$\sin \beta = e, \tag{A3.3}$$

while outside the spheroid

$$\sin \beta = ae/\tilde{\omega}. \tag{A3.4}$$

4. If a body is composed of concentric spheroidal shells of the same eccentricity e and if the density of a shell is $\rho(a)$, where a is its semi-major axis, the gravitational field in its midplane is

$$g_{\tilde{\omega}} = -4\pi G \sqrt{(1-e^2)} \int_0^{\tilde{\omega}} \frac{\rho(a)a^2 \mathrm{d}a}{\tilde{\omega}\sqrt{(\tilde{\omega}^2 - a^2e^2)}} = -G \int_0^{\tilde{\omega}} \frac{\mathrm{d}M(a)}{\tilde{\omega}\sqrt{(\tilde{\omega}^2 - a^2e^2)}}. \tag{A3.5}$$

The same formula is valid both inside and outside the body but the integrand vanishes for $a > a$, where a is the semi-major axis of the whole body.

Suggestions for further reading

Most books on astronomy contain chapters on galaxies or material that is relevant to this book. However, some of the recent developments in the subject, which have been mentioned, may not be covered in most recommended books.

Books which are descriptive rather than mathematical include:

B.J. Bok and P.F. Bok, *The Milky Way*, Harvard.

P.W. Hodge, *Galaxies*, Harvard.

P.W. Hodge (Editor), *The Universe of Galaxies*, Freeman.

E.P. Hubble, *The Realm of the Nebulae*, Dover.

W.J. Kaufmann III, *Galaxies and Quasars*, Freeman.

Hubble's book discusses the observations which led to the idea of an expanding Universe but, as mentioned in the text, his value for the scale and expansion rate of the Universe is now known to be seriously wrong.

Many fine photographs of galaxies and an outline of their classification can be found in:

A. Sandage, *The Hubble Atlas of Galaxies*, Carnegie Institute of Washington.

Another good collection of galaxy photographs is:

T. Ferris, *Galaxies*, Sierra Club.

The currently standard textbooks on the structure of galaxies are:

D. Mihalas and J. Binney, *Galactic Astronomy*, Freeman.

J. Binney and S. Tremaine, *Galactic Dynamics*, Princeton.

A good discussion of the physics of the interstellar medium can be found in:

J.E. Dyson and D.A. Williams, *The Physics of the Interstellar Medium*, Manchester University Press.

L. Spitzer Jr., *Physical Processes in the Interstellar Medium*, Wiley.

A good recent lecture series on the structure of our Galaxy but with some material on other galaxies is:

G.F. Gilmore, I.R. King and P.C. van der Kruit, *The Milky Way as a Galaxy*, University Science Books.

204

Some books which are concerned with cosmology include:

J.D. Barrow and J. Silk, *The Left Hand of Creation*, Basic Books.

G. Gamow, *The Creation of the Universe*, Mentor.

E.R. Harrison, *Cosmology: The Science of the Universe*, Cambridge University Press.

P.J.E. Peebles, *Physical Cosmology*, Princeton.

P.J.E. Peebles, *The Large Scale Structure of the Universe*, Princeton.

J. Silk, *The Big Bang*, Freeman.

S. Weinberg, *The First Three Minutes*, André Deutsch.

Gamow gives the original account of the hot big bang cosmology and Weinberg's exciting discussion brings it almost up to date. The books by Peebles are advanced texts. The existence and nature of hidden matter in the Universe is discussed by:

J.R. Gribbin and M.J. Rees, *Cosmic Coincidences*, Bantam Books. M. Riordan and D.N. Schramm, *The Shadows of Creation*, Freeman.

R.J. Tayler, *The Hidden Universe*, (Revised edition) John Wiley and Praxis Publishing.

The structure and evolution of stars is discussed by:

E. Böhm-Vitense, *Introduction to Stellar-Astrophysics. Vol.3: Stellar Structure and Evolution*, Cambridge University Press.

R. Kippenhahn and A. Weigert, *Stellar Structure and Evolution*, Springer.

R.J. Tayler, *The Stars: Their Structure and Evolution*, (2nd edition) Cambridge University Press.

My book is at a similar level to the present one. The other books are more advanced.

Index